Animal Factories

Animal

Jim Mason

A Herbert Michelman Book

Factories

and Peter Singer

CROWN PUBLISHERS, INC. NEW YORK

Unless otherwise noted, all photographs are from *The Foto Files,* New York.

Inquiries should be addressed to Crown Publishers, Inc., One Park Avenue, New York, New York 10016

Printed in the United States of America

Published simultaneously in Canada by General Publishing Company Limited

Library of Congress Cataloging in Publication Data

Mason, Jim.
  Animal factories.

  "A Herbert Michelman book."
  1. Livestock—United States. 2. Animal food.
3. Animal industry—United States. I. Singer, Peter,
joint author. II. Title.
SF51.M37   636.08'3   79-26423
ISBN: 0-517-53844X (cloth)
      0-517-538822 (paper)

10 9 8 7 6 5 4 3 2 1

Designed by Leonard Henderson

First edition

Books by Jim Mason
Animal Factories

Books by Peter Singer
Animal Factories
Democracy and Disobedience
Animal Liberation
Practical Ethics
Animal Rights and Human Obligations
(edited with Tom Regan)

CONTENTS

ACKNOWLEDGMENTS

As we look back on the six years since the beginning of this book, one person stands out for her contributions. Alice Herrington, President of Friends of Animals, brought the authors together in 1974 and suggested that the two of us write this book. We knew something of the problems with animal factories for Peter had written about them before, but Alice suggested that we look further; so we did. Our research and efforts would never have turned into this book, however, without the assistance and encouragement of the following people.

We owe special recognition to all of the farmers, farm workers, agricultural specialists, and others who generously lent their time, advice, and hospitality, making it possible for us to see for ourselves and photograph conditions on factory farms.

Jim Behrenholtz and Connie Salamone provided papers and information on environmental problems associated with animal agriculture.

Herbert Snyder, Secretary of the National Council of Public Land Users, gave us information about grazing and land use in the West.

Tom Smith of the Community Nutrition Institute informed us about some of the behind-the-scenes activities of industry special interest groups in Washington, D.C.

Peter Roberts and Thelma Knight of Compassion in World Farming and Kim Stallwood advised on European legislation governing care and housing of farm animals.

Joanne Bower of the Farm and Food Society set us straight about the nutritional content of food from animal factories and about environmental problems associated with them.

Alex Hershaft, Robin Hur, and Azadi Kudes reviewed technical information in various chapters and helped us verify it or find better information.

Jennifer Killian and Esther Mechler assisted by translating European legislation and other materials into English.

Ede Rothaus and Kristin Johnson offered ideas and suggestions about design, format, and the use of photography.

Heidi Lindy and Susan Markey typed our numerous drafts and revisions and made many improvements along the way.

Eileen Myerson, Leigh Manasevit, Rita Hall, Bruce Brown, and Joe Patton contributed by providing hospitality and space in which to do darkroom work, research, and other chores.

We owe special recognition to Patricia Curtis and Patrice Greanville for reviewing the final drafts; their comments led to a greatly improved text.

Finally, we owe thanks to the editors and staff at Crown Publishers for their support for this book, especially Peter Burford and Herbert Michelman, whose thoughtful editorial advice has substantially improved our final manuscript.

To all of the above, we are grateful; the responsibility for any errors and inaccuracies, however, is ours alone.

J.M.
P.S.

December 1979

INTRODUCTION

by Jim Mason

THE REALITY OF A MODERN ANIMAL factory stands in sharp contrast to the farm of our fantasies. The farm in our mind's eye is a pleasant, peaceful place where calves nuzzle their mothers in shady fields, pigs loaf in mudholes, and chickens scratch and scramble about barnyards. The farm animals in our fantasy live tranquil, easy lives. So we like to believe—because if they have fresh air, good food, exercise, and rest, we have more wholesome and delicious meat, eggs, and milk.

I remember this kind of farm. I was raised on one in Missouri in the forties and fifties. The animals on farms then were truly domestic, that is, they were a part of the farm household. Families lived from and cared for their animals. Through natural growth and reproduction, the animals produced enough to supply the household and bring in a small but steady flow of cash from sales in nearby markets.

Farms like the one of my childhood are rapidly being replaced by animal factories. Animals are reared in huge buildings, crowded in with cages stacked up like so many shipping crates. On the factory farms there are no pastures, no streams, no seasons, not even day and night. Animal-wise herdsmen and milkmaids have been replaced by automated feeders, computers, closed-circuit television, and vacuum pumps. Health and productivity come not from frolics in sunny meadows but from syringes and additive-laced feed.

I began learning about the trend to animal factories in the early 1970s. I knew that our routine uses of animals involved restrictions and manipulations that would be regarded as atrocities if practiced on humans. I began to realize that to make things easier on ourselves as we make use of animals, we also hide or deny the unpleasant reality of what we are doing.

The real nature of factory farming was indeed hard to believe. The contrast to my own farm experience was too strong; I assumed that there were only a few factories and that they were isolated examples. As I looked more deeply, I was overwhelmed by the awesome scale and pervasiveness of this new way of animal rearing. I was amazed how little the public knew about these drastic changes in the production of their food.

My efforts to find out more about factory farming were given a boost in 1974 when I met Peter Singer, who was then teaching philosophy at New

York University and writing his book *Animal Liberation*. The book documents humans' tyranny over other animals and explains why it is wrong. Together we decided to look more closely at the trend to animal factories and to write this book. At first my view of factory methods went little further than the animal welfare problems that Peter exposed in *Animal Liberation*. Before long, I learned that "confinement," or "intensive" farming as it is called in agricultural circles, raises many other problems as well: health, waste, dislocation of rural communities, pollution, and an overemphasis on technological solutions to problems arising from animal nature.

Research into any major activity of this sort requires a firsthand look. I subscribed to farming magazines and wrote to agricultural schools and states for information. I traveled about 14,000 miles, visiting farms and agricultural research centers from Canada to Nebraska to Georgia. I spoke with farmers, scientists, and others involved in the trend toward factory farms. The methods described here and the photographs shown are not intended to be typical of all American agriculture. There are still many traditional, diversified farms. But these descriptions and photographs do illustrate the trend in animal production toward greater concentration of animals, large-scale production, labor-saving machinery, growth acceleration, and other factory methods.

It should be clear that this book is not intended to be an attack on farmers and the people who run factory farms. They are as much victims of factory technology, although in a different way, as are the animals. No stranger to the economics and other realities of farm life, I am familiar with farm people and the feelings they have about making a living from their land. They are buffeted by economic developments that force them to try to push their animals to greater and greater productivity. It is a double tragedy that many farmers wind up losing their farm livelihood because the present trend culminates in a way of farming beyond their financial reach.

This book, then, is about a dominant trend in perhaps our most basic economic activity. For the past 9,000 years, farm animals have relieved humans from much of the toil of production, allowing us to devote time to civilization. Today, animal agriculture is our largest food industry and one that affects our lives daily. We must not be blind to the changes it undergoes—not only for our own good but for the good of nature around us.

No light, but rather darkness visible
Served only to discover sights of woe,
Regions of sorrow, doleful shades, where peace
And rest can never dwell, hope never comes
That comes to all; but torture without end . . .
                    —John Milton, *Paradise Lost*

FACTORY

Animal Factori

LIFE

What It's Like
to Be a Biomachine

The modern layer is, after all, only a very efficient converting machine, changing the raw material—feedstuffs—into the finished product—the egg—less, of course, maintenance requirements.
—Farmer and Stockbreeder, January 30, 1962

Forget the pig is an animal. Treat him just like a machine in a factory. Schedule treatments like you would lubrication. Breeding season like the first step in an assembly line. And marketing like the delivery of finished goods.
—J. Byrnes, "Raising Pigs by the
Calendar at Maplewood Farm," *Hog
Farm Management*, September 1976, p. 30

What we are really trying to do is modify the animal's environment for maximum profit.
—M. D. Hall, "Heating Systems for
Swine Buildings," *Hog Farm
Management, 1976 Pork Producers'
Planner*, December 1975, p. 16

IN THE YEARS BEFORE WORLD WAR II, farmers near large cities began to specialize in the year-round production of chickens to meet the demand of nearby markets for eggs and meat. The discovery of vitamins A and D had made it possible to raise large flocks indoors; when these vitamins were added to the birds' feed, sunlight and exercise were no longer necessary for proper growth and bone development. Large-scale production caught on. Farmers built bigger buildings for larger flocks and larger profits. But large flocks meant more labor, and the crowding produced a host of problems. In the poorly ventilated new buildings,

1

contagious diseases were rampant, and losses multiplied throughout the budding poultry industry. Entire flocks were wiped out. Nightmarish scenes occurred in large, crowded flocks: some birds would peck others to death and eat their remains. Many farmers were ruined and gave up farming; but the more persistent looked for solutions to these self-induced problems.

During the war years, demand for poultry was high. The boom in the chicken business attracted the attention of the largest feed and pharmaceutical companies, and they put their experts to work on the new problems associated with large-scale production. The breakthroughs to commercial production by factory methods began to come thick and fast. Burning off the tips of birds' beaks was found to reduce losses from pecking and cannibalism. An automatic debeaking machine was patented, and its use became routine within a few years. Newly developed hybrid corn made chickens put on weight faster. Automatic, chain-driven feeders ended the chore of carrying feed to the birds. Automatic fans, lights, and other labor-saving equipment followed. A mechanical chicken-plucker with whirling, rubber-fingered drums increased processing capacity while lowering labor costs.

The chicken itself, however, was not quite ready for mass production. In 1946, the Great Atlantic & Pacific Tea Company (now A&P) launched the "Chicken of Tomorrow" contest to find a strain of bird that could produce a broad-breasted carcass at a low feed cost. Within a few years, the prototype for today's fast-flesh broiler was developed. In the late 1940s, sulfa drugs and antibiotics were introduced as additives in chicken feed, for they held down disease losses and stimulated the chicken's rate of growth.

Bigger money was now being attracted to mechanized broiler production while bankruptcies and takeovers of small farmers, feed companies, and processors eliminated the old ways. These "shakeouts" occurred because the small producers and operators could not muster the capital that the factory mode of broiler production now needed. The end of World War II brought chaos to the growing industry through the loss of its biggest customer, the United States government. Conditions were ripe for those who had the money to take over production from the shakeouts at bargain prices. In the fifties and sixties, the major feed companies bought up most of these broiler operations, and major pharmaceutical companies began buying up breeding companies that were producing commercial chicken strains.[1]

The Cage Arrives

News of the success on the meat side of factory chicken production had spread to egg producers, and they tried the same methods. But one major new problem literally piled up: the confined, egg-laying hens produced tons of manure each week. Broiler producers had had the manure problem in their large flocks too, but their birds were in and out within twelve weeks and accumulations could be cleaned out between "crops." Egg farmers, however, kept their birds indoors for a year or more. They had to find a means of manure removal that would not disturb the hens or interfere with production. Producers discovered that they could confine layer hens in wire-mesh cages suspended over a trench to collect droppings. The manure pile could be cleaned out without bothering the hens above. At first, producers placed their birds one to each cage. When they found that birds were cheaper than wire and buildings, crowded cages in crowded houses became the rule. Crowding did mean that more hens died, but this cost was slight compared to the increased total egg output. Between 1955 and 1975, flock size in a typical egg factory rose from 20,000 to 80,000 birds per house.[2] In 1967, 44 percent of commercial layers were caged; today, about 90 percent of all egg production comes from caged birds in automated factory buildings.[3]

It took some time before anyone attempted to extend the mechanization practices of the poultry industry to larger farm animals. But poultry industry successes were an irresistible model for livestock experts. During the 1960s, they designed and built systems for pigs, cattles, and sheep that incorporated the principles of confinement, mass production, and automated feeding, watering, and ventilation. Instead of wire cages, which could not hold these heavier animals (although young pigs are now kept in wire cages on some farms), indoor pens and stalls were built over manure pits and fitted with slatted floors of concrete or steel planks spaced a fraction of an inch apart. This signaled the end of hand labor with pitchfork or shovel; now the farmer merely pumped out the pit every month or so.

As in any other industry, the fine details in methods and systems vary, but the basic processes are fairly standard.

Manufacturing Egg-Laying Machines

The modern chicken is a business creation; it owes its existence to the sterile laboratories of a few "primary breeders." About six companies develop the strains of chickens specialized for egg and broiler production. They sell breeding males and females to some 600 "multiplier" companies which, in turn, produce the chicks that go to egg and broiler farms.

At the multipliers, breeding flocks are kept in a long, low breeding house. Their eggs are placed in an incubator for nineteen days and then transferred to a hatcher. When the chicks break out, they encounter the poultry industry's mass-handling techniques for the first time.

If the hatchery is turning out "egg-type" birds, the first order of

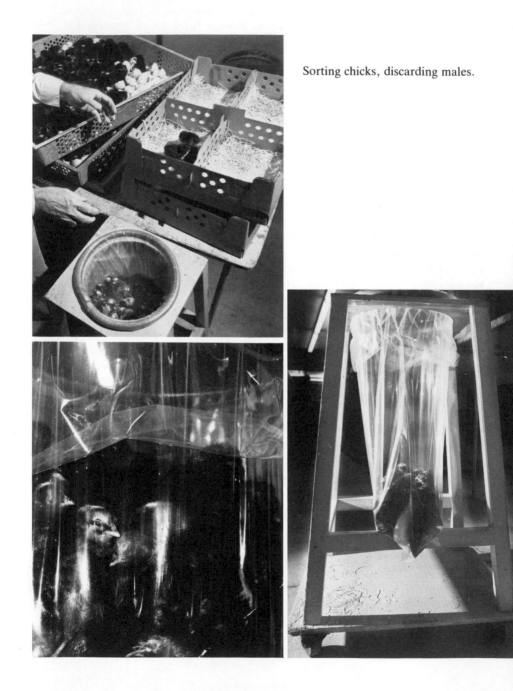

Sorting chicks, discarding males.

business is to destroy half of the newly hatched chicks. Males don't lay eggs, and the flesh of these strains is of poor quality. So they are, literally, thrown away. We watched at one hatchery as "chick-pullers" weeded males from each tray and dropped them into heavy-duty plastic bags. Our guide explained: "We put them in a bag and let them suffocate. A mink farmer picks them up and feeds them to his mink."[4]

Factory-Made Eggs

The female chicks are sold to egg producers to begin their careers as "layers." Egg producers clip the beaks and, in some operations, the toes of new chicks and vaccinate them against a variety of poultry diseases. Producers house the pullets (young hens) in "grow-out" buildings for about twenty weeks until they are ready to begin laying eggs. To hold down excitement and fighting as growing birds crowd the pullet house floor, many producers keep the buildings dark except at feeding time.

When the birds are mature, they are moved to the automated layer house where they are confined in cages that run the length of the building two, three, or four levels high. A watering system and two conveyor belts, one to bring feed and one to collect eggs, run along each level. The cage floor slants to allow eggs to roll through an opening onto a belt that carries them to a processing room for washing, grading, packing, and storage. Droppings from birds in the upper cages fall on sheet metal dropping boards mounted over lower cage rows. A power-driven winch and cable assembly moves a scraper along these boards, forcing the droppings into a pit below. When droppings accumulate in the pit, another power-driven scraper moves along under the row and pushes them out of the building.

In these specialized factory buildings, dotted around the United States, some 250 million hens produce virtually all our eggs. Although chickens kept under more natural conditions can live from fifteen to twenty years, hens in the commercial layer house last only about a year and a half. Their ability to produce eggs is diminished by the wear and tear of cage life until it becomes unprofitable to house and feed them. When this point is reached, they are made into soup and other processed foods.

Factory-Made Broilers

The early life of broiler chickens is similar to that of their cousins in the egg business. Here, however, males are kept and raised for market, although separately from females on many farms. Chicks are debeaked and toe-clipped at a day or two of age and taken to the broiler house. A partition restricts the young chicks under warm lights or gas heaters at one end of the floor. At first, bright lights are kept on most of the day to encourage the chicks to start feeding. After a week or two, the partition is

A modern egg factory.

removed and the young birds take over the entire floor. On most farms, the floor is covered with wood shavings or other litter material; on others, slatted floors—thin wooden strips spaced about half an inch apart—are built over a manure pit. Two or three rows of heaters, automatic feeders, and waterers stretch the length of the building.

The birds take about eight weeks to reach market weight—about 3½ pounds. During the last few weeks in the broiler house, when crowding is most severe, dim lights or near-darkness are used on many farms to reduce fighting. When the flock is ready for the slaughterhouse, the birds are crowded toward one end of the building. Then, at night when the birds

are drowsy, crews of "catchers" wade in, catch them, and load them in crates stacked on the trucks outside.

This reliance on human labor seems old-fashioned, and the broiler industry is trying to grow birds in cages similar to those used by egg producers. Birds could then be reared and shipped to market, cages and all. There are problems, though, in that caged broilers develop breast blisters, bruises, and foot and leg injuries. Undaunted, the primary breeding companies are trying to perfect a strain that can withstand the cages. Once they succeed, the broiler industry will switch to these birds and retool for more thoroughly mechanized production.

Poultry growing units will become larger and larger. Cage production of broilers will gradually take over. The new "factories" will move the birds to larger equipment. . . . A push-button control system will move the belts supporting the cages. The housing units will be on a flow-through basis—chicks will enter at one end and the finished live broiler will come out at the other.[5]

Systems such as this will eliminate the labor of catching the birds; moreover, cages can be stacked so producers will be able to house three or four times as many birds in the same space. There are other possibilities for the elimination of labor in the broiler business: a mechanical "harvester," like a giant vacuum, sucks birds from the floor and sends them through a large hose into crates. An English equipment manufacturer has designed broiler cages with a drop-away floor that dumps birds onto a conveyor belt.

More than 3 billion chickens are factory reared each year for American tables. Ducks and turkeys are mass produced by similar methods.

Factory-Made Pigs

Pig farming is concentrated in the Corn Belt close to the feed; 80 percent of U.S. pigs are raised in thirteen midwestern states.[6] The structure and methods of pig farming are rapidly changing. Although most pig farms are still family owned and operated, about 90 percent of pigs are in some type of confinement system.[7] The degree of confinement varies considerably. On many farms, pigs are born in confinement but grow up outdoors in pens or on pasture. The trend, however, is toward larger farms, factorylike mass-production methods, and a greater degree of restriction for farm animals. If it continues, pig production may soon become as automated and monopolistic as broiler production. The number of very large new farms is increasing rapidly.[8] The new large-volume farms adopt highly mechanized "total-confinement" systems. Pigs never see daylight until they are sent to market; they are conceived, born, weaned, and "finished" (fed for market) in specialized buildings boasting automatic feeding, watering, manure removal, environmental control, and other features similar to those in commercial poultry operations. Agricultural experts predict that by 1985, 65 to 70 percent of all pigs will be raised in total-confinement systems.[9]

The total-confinement pig farm is specially designed for maximum exploitation of the pig's reproductive and growth cycles. Typically, the farmer maintains a breeding herd of about 300 or more sows (females) and a few boars (males). These animals produce the "crops" of pigs that are fed and sent to market; they are not sold as long as their reproductive performance is satisfactory. Because breeding animals in controlled-

"The modern fowl thrives on a diet almost totally foreign to any food it ever found in nature. Its feed is a product of the laboratory."—Wilbur O. Wilson, "Poultry Production," *Scientific American,* July 1976, p. 58.

A modern pig "farm."

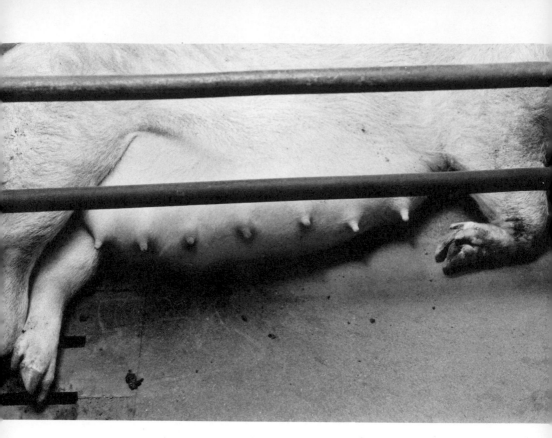

A sow in her farrowing stall.

environment buildings tend to lose interest in sex, most farmers move them outside at breeding time. Some use artificial insemination to eliminate the repetitive, time-consuming labor involved in moving animals to and from pens during breeding.

After conception, sows are put in a "gestation" building for about sixteen weeks. In this building, a sow shares a small pen with a few other pregnant sows, or, in newer systems, she is confined to an individual stall. In either event, she remains in her pen or stall for the entire sixteen weeks—all, so the experts say, to give the producer "better control" over pregnant sows. In fact, the strict confinement creates a need for controls: to hold down stress and excitement, the rooms are kept dark except at feeding time; to hold down excessive weight gain (and feed bills) in the inactive sows, they are "limit fed" once every two or three days.

About a week before her piglets are due, the pregnant sow is moved to a "farrowing" building to give her time to settle down and adjust to the change before the births. Here she is restricted to the tight quarters of a farrowing stall for two or three weeks until her piglets are born, nursing, and on their feet. The stall permits her to lie and stand, but she cannot walk or turn around; its purpose is to keep her in position only to eat, drink, and keep her teats exposed to the young pigs.

The newborn pigs receive injections of iron and antibiotics, their "needle" teeth are clipped, their tails removed, and their ears notched for identification. The young males are usually castrated just before weaning. Two or three weeks after birth, the sow goes back to the breeding area and her pigs are taken to other buildings to start being fattened for market.

In the "growing" buildings, the pigs encounter the usual factory equipment that removes manure and provides them with a steady supply of ground corn and other nutrients. They remain here for about twenty weeks until they are ready for market, at about 220 pounds. Farmers have now begun to stack pens and cages two and three decks high to expand production in the growing buildings while saving space and labor.

Confinement systems for pigs come in all shapes and sizes. A few producers use ingenuity and available materials to put up systems of their own design, but most choose from systems available from some sixty confinement manufacturers. The pages of farming magazines contain advertisements for "push-button farming," "plug-up-and-go systems," and "totally engineered" systems designed to tend the pigs while leaving the farmer only the job of checking on the equipment now and then. An Iowa pig producer and owner of a 400-pig nursery and finishing building says, "It's just a matter of five minutes in the morning, and we try to check them every evening." [10] One manufacturer of the new systems advertises that with its systems "just a turn of the control dials in better pork profits." [11]

Factory-Made Milk

Both meat and dairy cattle are now being brought from pastures to mechanized confinement buildings as farmers turn to the new factory methods. On the factory dairy farm, cows live by the farmer's daily milk collection routine. For most of the day, a cow is confined to a holding area where feed and water are available. Twice each day she goes into a milking parlor where a system of rubber cups, plastic tubes, and vacuum pumps extracts her milk and pumps it into a refrigerated tank. If a profitable operation is to be maintained, a dairy farmer must keep each cow in the herd producing milk for as many days as possible throughout the year. Since cows give milk only for about ten months after the birth of a calf, farmers rebreed each cow soon—usually about forty to sixty days—after each calf is born.

The calf is taken from its mother when a few days old, for otherwise it would drink the milk that is intended for sale and for human consumption.

Cows awaiting milking.

If female, the calf may be raised for the herd; if male, it will be sent to market as soon as possible. After a few years of routine calf and milk production, the cow's milk productivity wanes, and she is sent to the slaughterhouse. Because her carcass is not good enough to produce steaks, chops, and other cuts, she will probably end up in the hamburgers of one of the fast-food chains.

Factory systems and methods came gradually to dairying several years ago. As land values rose and farm labor became more and more scarce, some farmers began to install mechanical milking systems that could ensure speedy milk collection from larger herds. Dairy cows, like chickens, had been reliable producers of spending money for small, mixed-crop farms; now they have been converted to mass producers of profits for specialized dairy farms. About half of all dairy cows and heifers today are kept in some type of confinement system.[12]

A total-confinement dairy setup consists of only two buildings: the mechanized milking parlor and the holding barn. On a large dairy farm, where several hundred cows go through twice daily, these milking parlors are highly mechanized. Electronic sensors, automatic gates, automatic feeders, and vacuum-operated milking machines ensure a steady flow of cows and milk. About all the operator has to do is to wipe each cow's udder with disinfectant, apply suction cups to the cow's teats, and watch the machinery. To keep the herd conveniently near the milking parlor, most of the large farms have holding barns attached, equipped with the usual mechanized feeding, watering, and manure removal gear. In "free-stall" holding barns, cows are free to move from their stalls and walk about concrete or slatted floors when not at the milking parlor. Other farmers prefer to keep their cows stationary in "tie-stall" barns and move portable milking machinery to them. These cows remain in stalls for months, chained at their necks.

Factory-Made Veal

Veal factories are the harshest confinement systems. Newly born calves are taken from their mothers and turned into anemic, neurotic animals to produce the luxury "white," "milk-fed," or "prime" veal preferred by gourmet cooks and fancy restaurants. Compared with the production of other animal commodities, milk-fed veal production is a small industry. Only a few hundred thousand calves go through these factories each year.[13]

Because they require an ample supply of newborn calves, veal factories tend to be located in dairying regions. The veal producer buys day-old calves and places them in individual stalls in the veal factory building. For fourteen to sixteen weeks the calf is confined to a space scarcely larger and wider than its own body, often tied at the neck to further restrict movement within the stall. Throughout this confinement the calf is fed on "milk replacer," a mixture of dried skim milk, dried whey, starch, fats, sugar, mold inhibitors, vitamins, and antibiotics. Commercially made replacers are high in fat to cause the immobile calves to gain weight rapidly.

A visit to a milk-fed veal factory in northern Connecticut gave us a feel for the business of veal production. Although it was broad daylight outside, the calves' rooms were pitch-dark; our guide explained that darkness helped keep the calves quieter.[14] At feeding time the lights were turned on as the producer made his rounds. In two rooms, more than a hundred calves were crated in rows of wooden stalls. Their eyes followed our movements; some appeared jittery, others lethargic. Many tried to stretch toward us from their stalls in an attempt to suckle a finger, a hand, or part of our clothing. The farmer explained: "They want their mothers,

Veal factory.

I guess."[15] The farmer mixed milk-replacer in a barrel and rolled it along the rows of stalls. At each stall, he took a clean plastic pail, half-filled with the mixture, and hung it on the stall. This liquid, twice daily, was the only "food" allowed to the calves. Buckets with nipples could relieve the young calves' urge to suckle, but are not used because they have more parts and recesses that might harbor bacteria. Moreover, the disassembly, cleaning, and sterilizing of 200 buckets twice each day would be costly and time-consuming to the farmer. The stalls contained no straw or other bedding, for that could be eaten by the calves and the iron it contains would darken their flesh.

This goes on for fifteen weeks, by which time the calves weigh an average of 330 pounds and are ready for market. Producers would like them to gain more, but by this time the anemic condition is severe and the longer they are kept, the more will sicken and die in the stalls.

Factory-Made Beef

During the "beef crisis" in 1973, many Americans became critical of our wasteful and inefficient practice of feeding grain to cattle for the production of tender, fat-marbled steaks. Beef industry promotional organizations responded that cattle produce protein from grasses and roughage not edible by humans. They did not tell us, however, that more than 14 million cattle and calves were then being fattened on soybeans and on corn, barley, and other grains.[16] Whatever industry propaganda says, most cattle marketed are still grain-fed in feedlots at some point before slaughter. Although the use of feedlots and confinement systems has fallen slightly, they are still very much with us. U.S. Department of Agriculture estimates for 1976 list 134,417 feedlots in the United States, most of them in the north-central states and Kansas.[17] Of these lots, 1,750 large operations marketed two thirds of the total production and 60 "super" lots turned out 18 percent of all fed cattle in that year.[18]

On conventional beef farms and ranches, producers maintain herds of "brood" cows and bulls. These "cow-calf" operations are more likely to be located in southern and western states where pastures and roughage are more available than corn and other feed grains. On these farms, calves are allowed to remain with the herd on pasture or rangeland until they are sent to market at one to two years of age. Some farmers round up yearling calves, confine them, and grain-feed them for three or four months before sending them to market; others sell "feeder calves" four to six months old to be fattened by their purchasers.

For many beef producers, however, these operations require too much land and labor and the calf "crops" come too slowly and too far apart. These are the cattle feeders whose speedier methods of beef production now dominate the beef industry. They buy feeder calves, put them in a feedlot, and finish them as quickly as possible to a market weight of about 1,000 pounds. In combination with diethylstilbestrol,* estradiol, progesterone, and other growth promotants, the cattle are fed high-calorie feed consisting chiefly of ground or flaked corn and soy meal or another protein supplement. The feedlot itself is typically a fenced area with a concrete feed trough along one side. In regions where land is expensive and grain is cheap, cattle feeders use total-confinement buildings with automatic feeding and waste removal systems similar to those in other animal factories.

---

*On November 1, 1979, the use of DES in food animals became illegal. See text, page 69.

A large cattle feedlot, Texas.

The Frontiers of Factory Farming

Other species of animals are now being reared in factory systems. Sheep experts at the University of Illinois have developed model confinement buildings in which growing lambs can be penned and fed to market weights. Sheep experts at Ohio State University have designed and built a model confinement unit that incorporates floors made of heavy-duty steel mesh, electrically powered manure removal, and automatic feeding by electric timer. All of this fits in with the American Sheep Producer Council's "National Blueprint for Expansion," designed to revive the sheep industry and bring it to eastern states.[19]

USDA photo

In cage systems similar to those used in egg factories, domestic rabbits are being raised for the meat and laboratory animal markets. Even the earth's waters are under invasion for more species to be converted into commodities by factory methods. In the Bahamas, green sea turtles are being factory-farmed in concrete and fiberglass tanks to produce steaks, soup, leather, and jewelry. In Florida and Louisiana, alligators are mass produced in confinement for the handbag and shoe trade. And although the subject is beyond the scope of this book, millions of fur-bearing animals such as mink are confinement-reared each year to supply the ever-increasing demand for fur garments.

Regardless of the type of animal confined or the commodity purchased, all factory systems are designed to make more money from more animals.

Instead of hired hands, the factory farmer employs pumps, fans, switches, slatted or wire floors, and automatic feeding and watering hardware. The factory farmer is a capital-intensive farmer whose greatest investment is in time- and labor-saving equipment. Success in farming is not achieved by direct care for the animals. It does not depend on the well-being of individual animals or even on individual productivity. Success comes from maximally efficient use of equipment. It is measured by year-end production records. Like managers of other factories, capital-intensive farmers are principally concerned with cost of input and volume of output. A certain amount of wastage doesn't matter if the product wasted is cheap by comparison with overheads and if eliminating the wastage would raise costs or reduce output. All this is as true of animal factories as of any other factory; the difference is that in animal factories the product is a living creature capable of pain and fear, a creature worthy of moral consideration that inanimate objects neither require nor could benefit from.

Rabbit factory.

Estrus control will open the doors to factory hog production. Control of female cycles is the missing link to the assembly-line approach.
—Earl Ainsworth, "Revolution in
Livestock Breeding on the Way,"
Farm Journal, January 1976, p. 36

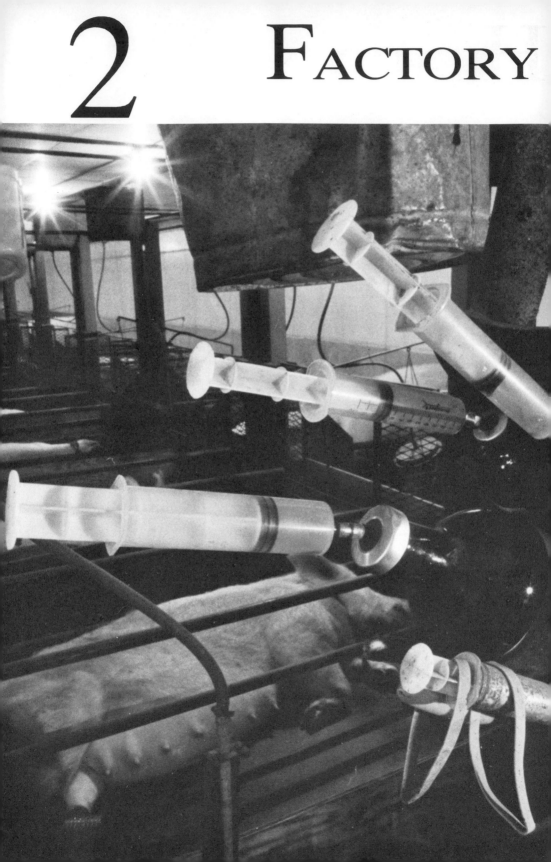

HEALTH

Even Biomachines Get Sick and Die

Some have said that with our growing management sophistication and heavy concentration of animals in small areas, there's a danger of some entirely new disease popping up—not unlike the Andromeda Strain *in science fiction.*
—"Can We Keep Our Livestock Healthy?" *Farm Journal,* mid-March 1978, p. Beef-21

The more complicated animal production becomes, the more effort you have to put into controlling some of these [disease] problems.
—Bob Glock, D.V.M., Iowa State University, Department of Veterinary Pathology, quoted in "Disease Threats Demand Total Health Program," *Hog Farm Management,* September 1977, p. 66

ENVIRONMENTS FOR FACTORY ANIMALS are designed for efficiency, not comfort. Barren cages and stalls in darkened rooms do not satisfy even the stunted psyche of a factory animal. These animals are bored, frustrated, and fearful—as animal scientists say, "stressed." Until recently, no animal scientist would gamble his or her respectability by examining how animals *felt* about their living conditions. Such a study would imply anthropomorphism, a cardinal sin for a scientist, who is supposed to record only observable data. Lately, however, a few progressives have broken the taboo because, as they say, "A much more detailed understanding of the animals' psychological as well as physiological requirements (while also considering economic controls) would lead to greater returns."[1]

Stress and Boredom

Factory animal experts are beginning, then, to look at stress, and for good reasons. In reacting to stress, animals burn up available energy and nutrients that would otherwise go toward growth, gestation, lactation, and disease resistance. The blood of a stressed animal carries increased levels

21

After months of rubbing against the wire, hens lose their feathers.

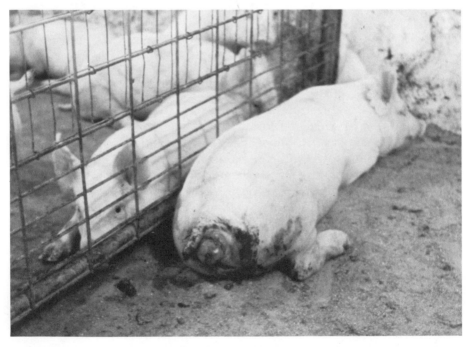

Victim of tail biting.

"The longer sows are kept in confinement the more problems occur. In general . . . sows kept under tethered conditions have done the poorest."—Neal Black, "Production Drops If Sows Confined," *National Hog Farmer,* November 1974.

of adrenal hormones that can break down muscle tissue into energy to meet the cause of stress. When an animal is stressed, its defenses are down and it is more prone to infectious diseases. Animals respond to stress in varying degrees; one particular animal may be stressed by a condition that produces no reaction in its penmate. But the stressed animal may catch a disease which spreads to the second animal.[2]

In the factory, animals are subjected to a variety of stresses. When birds are debeaked or when calves or pigs are weaned prematurely, some die from the shock. These causes of stress are occasional, however, and after a few days of adjustment most of the animals return to "normal." But other causes of stress on the factory farm are continuous: the animals have no relief from crowding and monotony. In a less restrictive environment they would relieve boredom by moving; confined animals cannot. Their instinctive drives are twisted into abnormal behavior.

When animals are crowded and annoyed, the likelihood and frequency of aggressive encounters increases. When growing pigs are moved to larger pens, outbreaks of fighting can occur, leaving pigs dead or injured. In the restricted space of confinement pens, less aggressive animals cannot get away to make the show of submission dictated by instinct. Some animals may become so fearful that they dare not move, even to eat or drink. They become runts and die. Others remain in constant, panicked motion, a neurotic perversion of their instinct to escape. Cannibalism is common in swine and poultry operations. Cannibalism in poultry results from a distortion of the birds' instinct to establish a social hierarchy, or "pecking order." Birds that have evolved over millions of years, socializing in flocks of about a hundred members, cannot establish a pecking order among the thousands on the floor of a modern broiler or turkey house. In these superflocks, birds would peck each other relentlessly if controls were not used. Caged birds have the opposite problem: each cage contains a small "flock" and somebody has to fall at the bottom of the social ladder. This unfortunate bird cannot escape its tormentors.

In pigs, cannibalism takes the form of tail biting, described by one expert as follows:

> Acute tail-biting is often called cannibalism and frequently results in crippling, mutilation and death. . . . Many times the tail is bitten first and then the attacking pig or pigs continue to eat further into the back. If the situation is not attended to, the pig will die and be eaten.[3]

Heightened levels of aggression and activity take their toll of stressed animals in a more direct way. Like any overworked machine, they simply wear out. Pigs in particular are prone to a reaction that we would probably call "shock" if it occurred in humans: the pig industry calls it PSS or the Porcine Stress Syndrome. Pigs may literally drop dead from stress when they are weaned, moved to a new pen, mixed with strange pigs, or shipped to market.

Reproductive functions are not essential to survival at the moment of stress, and so are pushed into the background until the animal has escaped the stress-causing situation. Hence confined male pigs lose sex drive, females fail to become pregnant, and the daughters of these animals may have incompletely developed reproductive organs and may be slower to reach puberty.[4] Similar disorders of reproductive behavior occur in other confined animals.[5]

Crowding and stress can disrupt normal patterns of behavior between parents and offspring. They may be unable to recognize each other because of the confusion of smells, sounds, and visual stimuli in the factory. In the farrowing house, for example, a bored, detached sow may not recognize the squeals of one of her own litter should she roll over or step on one. (On several occasions in the farrowing houses we visited, we saw young pigs trapped under their mothers' legs or rumps, squealing and struggling to free themselves. One farmer told us that he lost pigs this way even though his restrictive farrowing stalls were supposed to prevent sows from crushing their pigs.) Stress and stimulus confusion can also cause mothers to abandon their offspring and to refuse to accept their suckling, or the same conditions can make the young animals unable to seek out their mothers.

If you've been to a zoo and seen an animal repetitively rock back and forth or pace up and down its cage, you've seen another type of abnormal behavior, known as stereotypic behavior. It is a response to a boring environment. On the factory farm, sheep, calves, and pigs bite the bars of their stalls and caged chickens preen and pace. At one pig farm we visited, we watched one young pig methodically bite the large-gauge wire on the gate of her nursery pen. Starting at one side of the gate, she would bite at each square, work her way across, and then continue back and forth. Occasionally she would break this routine to run to the rear of her pen, then return and begin the pattern all over again.

Diseases in the Factory

The "controlled environment" of the animal factory can be a hothouse of air pollution and airborne germs. Even with powerful ventilators working properly, the air of pig and poultry factories contains dust raised by mechanical feeders and excited animals. It is also full of ammonia and other irritating gasses from the manure pits. Because factory buildings are usually in use year-round and isolated from the cleansing effects of sunlight and rain, they develop what producers call "bacteria buildup." A producer may have relatively few health problems in a new factory building during the first year or two, but eventually the interior can become infested with a variety of disease-causing organisms. Farming magazines indicate that both pig and dairy factories, for example, are plagued with diseases, some of which are aggravated by factory conditions.[6]

Confined calves suffer perhaps the worst disease problems of all factory animals. According to *Successful Farming*, "Calf losses are 15–20% on the average dairy farm. In some cases they zoom up to 40–50%."[7] In both veal and dairy operations, all of the conditions are right for disease: the animals are very young and highly stressed by separation from their mothers, their diet is inadequate, and they are concentrated indoors and unable to exercise. Veal calves' iron-deficient diet results in progressive anemia and marked stress; consequently they are more vulnerable to infections. The most common causes of death in veal factories are pneumonia and acute diarrhea.

Factory farmers' attempts to control some diseases often aggravate others. In more traditional systems of poultry farming, flocks picked up immunities from gradual exposure to germs harbored by their parents and older birds, and from the housing. Many modern flocks, however, have been bred to be "specific-pathogen free" (SPF), that is, certain disease-causing agents have been eliminated from the flock by breeding and raising birds under sterile conditions. The isolation necessary to successfully produce SPF chicks causes them to be isolated from other pathogens as well, and as a result they lack immunity to even ordinary microorganisms. When these birds are placed in the poultry buildings, they may develop "diseases never before observed by pathologists."[8]

The factory operator, if a good manager, tries to control temperature, humidity, light, ventilation, drafts, dust, odors, noise, fighting, diseases, waste removal, the supply of food and water, and everything else that makes up an animal's environment. But when hundreds or thousands of animals are confined in a single room, it is unlikely that every element of the environment is satisfactory to every individual animal. Health then suffers, and the causes are so diffuse that they are difficult to trace.[9]

Some producers attempt to maintain near-hospital conditions in an effort to reduce disease problems. Many factory farms do not allow visitors. At one of the large pig factories we visited, we were required to scrub, shower, put on white coveralls, and spray our photographic gear with disinfectant. On many farms, we waded through disinfectant footbaths or pulled thin polyethylene coverings over our shoes and pants.

As mentioned briefly earlier, the anemic condition deliberately induced in calves to produce high-priced, pale veal is one of the most severe of the

conditions suffered by factory animals. The relevance of anemia control to white flesh production is explained in an excerpt from a pamphlet, *Raising Veal Calves,* published by the University of Massachusetts.

> The anemic condition resulting from milk-only feeding gives the desired lack of pigmentation to muscle tissues in the carcass. Milk is deficient in iron and thus a food source that will give the desired end product. Other feeds must be avoided, as they do contain iron. Trace mineral salts, hay, bedding are excellent sources of iron and must be avoided. Grains should also be avoided in the feeding program.
>
> Physiologically, this type of nutrition program results in marked stresses to the animals, especially as they approach ten weeks of age. . . . The calf does have stores of iron in the body at birth, primarily in the form of extra hemoglobin in the blood, with lesser added amounts stored in the liver, spleen and bone marrow . . . [that] can carry the calf pretty well through 8 to 10 weeks. However, by this time the calf is anemic, and with a longer growing period, there is a decreased energy utilization, decreased resistance to disease, decreased growth rate, and eventually death.[10]

Barbed wire and chain-link fences: not to keep animals in but to keep germ-bearing visitors out.

Dead calves.

One producer explained to us that restriction to stalls assists the development of pale flesh by keeping the calves from "running around a little bit too much." He added that "keeping them confined does not allow muscle development and it keeps the meat tender."[11] When they are a few weeks old, the calves develop an urge to ruminate, that is, to eat roughage such as hay which is regurgitated and chewed as the "cud." In their craving for iron and roughage, veal calves attempt to chew their stalls and will lick at nails or any metal about them. Not surprisingly, digestive

The health of the veal calf can best be described as anemic, weak and susceptible to disease.

—Stanley N. Gaunt and Roger M.
Harrington, eds., *Raising Veal Calves*
(n.p.: The Massachusetts
Cooperative Extension Service, n.d.),
no. 1066, p.4

disorders, including ulcers and chronic watery bowel movements, are common in these calves.

Factory Diets

Even when not deliberately deficient in an essential nutrient like iron, the factory farm animals' feeds are designed primarily for cheap weight gain and consequently nutritional deficiencies do occur. Even if feed is properly formulated (mixing errors do happen), some animals get inadequate diets. At the feed mill, nutrients are added in amounts according to what the "average" animal needs. Because of stress or individual differences, some animals need more of an essential nutrient than they get. But if prices are right, a large factory still makes money even if large numbers of animals die or suffer from dietary deficiencies. Vitamin deficiencies common in poultry factories, for example, result in a variety of conditions, including retarded growth, eye damage, blindness, lethargy, kidney damage, disturbed sexual development, bone and muscle weakness, brain damage, paralysis, internal bleeding, anemia, and deformed beaks and joints.[12]

Dietary deficiencies and other factory conditions can cause a variety of bodily deformities. In poultry, fragile bones, slipped tendons, twisted lower legs, and swollen joints are among the symptoms of mineral-deficient diets. Birds raised on wire-mesh floors tend to develop crooked breast bones. Some poultry diseases can leave birds with malformed backbones, twisted necks, and inflamed joints. In broiler chickens, bruised and swollen hocks are common because the industry has produced a sluggish, top-heavy bird that spends most of its time "down on its haunches."[13]

Digestive disorders; nutritional deficiencies, and other health problems in cattle are on the rise now that more and more of these animals have been moved from pasture to feedlots and factory dairy systems where they are fed primarily grain. Cattle have a digestive system equipped to handle a diet of grasses, stalks, stems, and other roughage. The first of the cow's four stomachs, the rumen, contains populations of one-celled plants and animals that help break down the tough plant stalks and fibers. A diet of high-energy grain disturbs the proper environment for these microorganisms and several side effects occur: the lining of the rumen may become damaged, protozoa are destroyed, and the types of bacteria change. Certain abscess-producing bacteria proliferate, penetrate the rumen, enter the bloodstream, and settle in the liver. This condition, said to occur in most every feedlot, is termed "acidosis-ruminitis-laminitis-liver-abscess syndrome." It is responsible for 90 percent of all the rejections of beef liver by meat inspectors at slaughterhouses.[14] Despite this maldigestion, beef producers continue to feed grain because it produces a heavier animal more quickly and consumers prefer tender, grain-fed flesh to that of less intensively fed animals.

In factory dairy systems, new health problems are traceable, in part, to deficiencies of vitamins A, D, and E, which are available in forage and pasture diets but are not adequate in the premixed feed of the confinement barn.[15] Predictably, these vitamin deficiencies are corrected not by turning the cows out to pasture but by resorting to injectable vitamins and feed additives.

Factory Feet and Legs

The feet of birds and mammals contain complex joints composed of numerous small bones, ligaments, cartilage pads, tendons, and muscles. This complexity of design is not gratuitous; it equips the animals to scratch for food, kick or claw for defense, and stand or move on different types of terrain. About the only kinds of surfaces for which these feet and legs are not designed are the wire-mesh, concrete-slab, and metal-slat floors of confinement buildings. On these alien surfaces, some joints, muscles, and tendons are overworked, others are underworked, and consequently the complex structure of the animals' feet breaks down. Pigs are cloven-hoofed animals and in most the outer half of the hoof ("claw") is longer than the inner half. Outdoors, this extra length is absorbed by the natural softness of soil. On the concrete or metal floors of the factory pen, however, only the tissue in the foot can "give." As a result, many confined pigs develop painful lesions in their feet which can open and become infected. Pigs with these foot sores usually develop an abnormal gait and posture in an attempt to relieve the pain. Eventually the crippling may worsen when this abnormal movement and weight distribution overworks joints and muscles in the legs, back, and other parts of the pig.[16]

Foot and leg weakness has been accidentally bred into modern animals as breeders have turned out heavier-bodied, "meat-type" animals; their old-fashioned feet and legs can't take the extra strain.[17] Lameness, observed in bigger, faster-growing broiler birds. This condition "is characterized by birds jumping into the air, sometimes emitting a loud squawk and then falling over dead." [19] Upon post mortem examination, the bird's heart is full of blood clots, though these are believed to be a result rather than a cause of death. At one of the broiler operations we visited in North Carolina, the operator had been losing several birds each day from this condition, which he called "heart attack." He told us that the problem is "in the birds—they grow too fast these days." [20]

Factory Hazards

Under natural conditions an animal's instincts, conditioning, and senses guide it to food, steer it from danger, and generally promote survival. Factory production strips away the animal's freedom and ability to take care of itself. Control of life-support activities passes from the animal to machines and managers. In this vulnerable situation a power failure, a slight maladjustment, or a minor mistake can quickly cause the death of a very large number of animals.

Factory-farmed animals are vulnerable to less obvious hazards as well. Young birds die of thirst because they do not learn to drink from nipple-type watering devices; after debeaking, they can starve to death within inches of an adequate supply of feed. [21] Because of the large concentrations of animals in factory systems, lack of adequate ventilation can be critical, especially in hot weather. Metabolic activities produce heat and a "hot-air envelope" develops around the body of each animal. If the temperature becomes too high, the animals may just "plop over and die." [22]

The wave of health problems associated with the industrialization of animal agriculture is not abating. Industry leaders speak glowingly of eliminating diseases, but they succeed only in making disease control more complicated. The solution is to abandon the new methods of concentrating and confining animals; but economic pressures on farmers, together with a pro-confinement bias on the part of many of the industries that sell drugs, equipment, and other products to farmers, are hastening the conversion of animal agriculture to factory systems.

And what do we hear from the veterinary profession about the health problems of animals in factory systems? A recommendation that large-scale confinement systems be eliminated?

No. At present, factory farming pays and it pays handsomely. It pays for drugs, medicines, feed additives, and veterinary services. In any profession, when ethics and economics collide, ethics has a hard time of it. Nowhere is the collision more head-on than in factory farming, where the animals are the patients but the farmers pay the bills. Some veterinarians

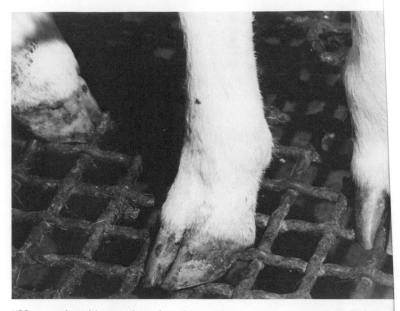

"Unquestionably modern housing management systems result in m[...]
stress to feet and legs than preconfinement production."—Robert D.
"Floors and Their Effects on Feet and Leg Problems in Swine," Co[...]
June 1976, p. 12.

however, affects only the living animal, not its flesh. Afflicted ani[...]
valuable meat whether or not they can walk, and they usually g[...]
slaughterhouse before they can become a liability to the farmer.

Caged birds often develop foot and leg problems as a result of[...]
of standing on a wire-mesh surface. To allow droppings to pass[...]
the cage floor with the least interference, cages are manufacture[...]
rather large-gauge mesh—usually about an inch square; thus the bi[...]
stand on only a few fine wires, which is an unsuitable perch for [...]
bird.

Worn-Out, Flipped-Out Biomachines

Factory experts are puzzled by some health problems in factory [...]
for which they are unable to find precise causes. One condition c[...]
in layer operations is termed CLF, or Caged Layer Fatigue. Accor[...]
Poultry Digest, birds with CLF withdraw minerals from their bon[...]
muscles and eventually are unable to stand.[18] These fatigued bird[...]
brittle or broken bones and a pale, washed-out appearance in thei[...]
combs, beaks, and feet. Another mystery is the "flip-over synd[...]

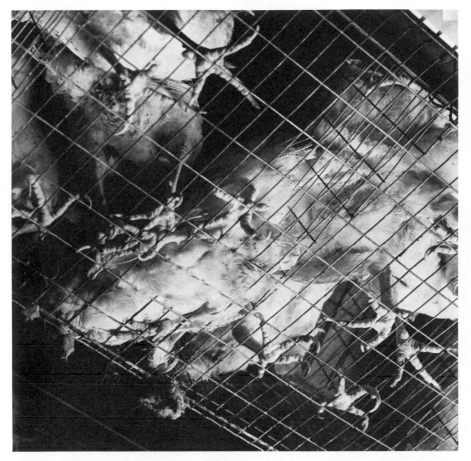

"Birds get their legs caught in the wire floor of the cage and die because they cannot get to feed and water."—C. I. Angstrom, "Mechanical Failures Plague Cage-Layers," *Onondaga County Farm News* (Syracuse, New York), December 1970, p. 13.

may boldly criticize factory farming, but the majority will go on trying Band-Aid solutions: more drugs, stricter isolation from the natural environment, better equipment, and a reformulated diet. We cannot look to the veterinary profession for the fundamentally different approach that is needed.

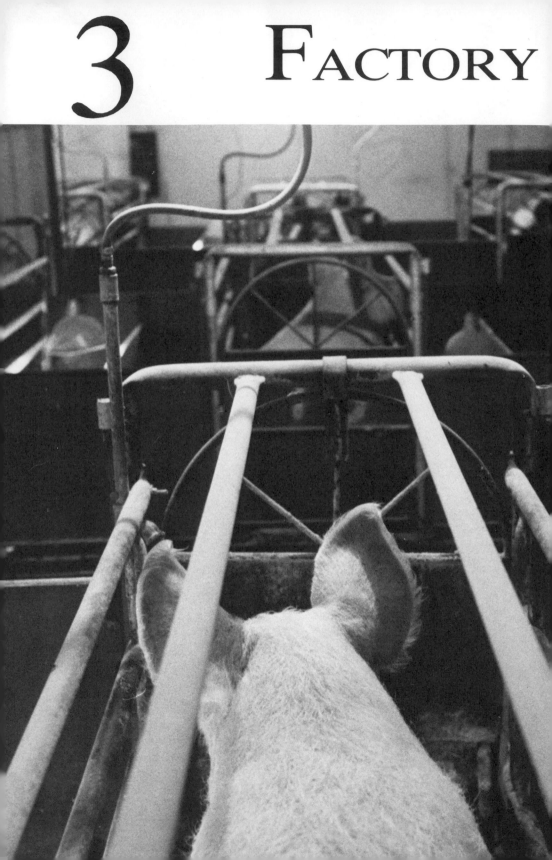

MANAGEMENT

The Biomachine Designed, Assembled, Modified, Tuned In, and Turned Up

The breeding sow should be thought of, and treated as, a valuable piece of machinery whose function is to pump out baby pigs like a sausage machine.
—L. J. Taylor, export development manager for The Wall's Meat Company, Ltd., in *National Hog Farmer,* March 1978, p. 27

At the Animal Research Institute (Agriculture Canada) we are trying to breed animals without legs and chickens without feathers.
—R. S. Gowe, Director of the Animal Research Institute, Agriculture Canada, speaking at "Livestock-Intensive Methods of Production," a conference in Ottawa, December 6 and 7, 1978

THE NEW PROBLEMS OF FACTORY FARMING are seen as "challenges" for an ever-growing army of experts who churn out increasingly elaborate management schemes to keep the factory profitable. Despite the claims of equipment manufacturers advertising "push-button farming," the new systems need careful management. Constant manipulations of animals' anatomy, physiology, heredity, and environment are required to keep health problems and other costs down so that commodity production can proceed at a profitable level.

Farmers know that crowding, diet, air pollution, and other environmental conditions aggravate their animals' health problems, but because elimination of these adverse conditions would require elimination of factory techniques, farmers turn to those few manipulations that can be carried out within the factory system. Stress and related health problems can be reduced by keeping animals in darkness or under very-low-intensity lights. The total-confinement veal and pig operations we visited kept animals in total darkness around the clock except for brief inspection and feeding periods. Egg factories cannot be kept dark around the clock as the illusion of day is necessary to stimulate egg production. But research under way at Cornell University may change that. Scientists there believe that hens need only brief exposures to light at critical moments during their ovulation cycle.[1] Once these moments are mapped out, layers should lay as usual with a total of only two or three hours of light each day.

Chemistry, too, is used in attempts to control stress. Poultry scientists at Southern Illinois University found that an antibiotic feed additive "apparently relieved the stress of crowding" in layer chickens.[2] Researchers at Virginia Polytechnic Institute suggest that poultry producers may soon be able to "use steroid drugs to block stress factors that interfere with optimum performance."[3] Research of this kind is good news to drug and chemical companies that are only too happy to market these problem-solving additives. For stressed pigs, a company sells an additive for feed claimed to "stop tail-biting within 40 hours."[4] Researchers are studying the possibility of using other chemical feed additives to control tail biting.[5] While results so far are inconclusive, research of this type indicates the thinking of many factory experts: ignore crowding and other causes of stress, just look for ways to mask symptoms that affect production.

Stressful situations can be greatly reduced by feeding chemicals such as metyrapone or DDD (rhothane) which reduce the production of corticosterone by the adrenals.

—Paul Siegel and Bernie Gross,
"We're Learning How to Let Bird
Defend Itself," *Broiler Industry,*
August 1977, p. 42

Cutting Off the Offending Part

Manipulations of environment and chemistry may lower stress to some extent, but in pig and poultry operations additional steps must be taken to hold down losses from cannibalism. Management of factory farms calls for deliberate animal engineering: if the factory cannot be modified to suit the animal, the animal is modified to suit the factory. To ensure that stressed pigs cannot tail-bite, farmers routinely cut off ("dock") the tails of young pigs a few days after birth. In poultry, cannibalism is controlled by routine debeaking. This began around 1940, when a San Diego poultry farmer found that if he burned away the upper beaks of his chickens with a blowtorch, they were unable to pick and pull at each others' feathers. His neighbor caught on to the idea, but used a modified soldering iron instead. A couple of years later a local company began to manufacture the "Debeaker," a machine that sliced off the tips of birds' beaks with a hot blade. With various modifications, this machine still debeaks most factory birds. Broiler chicks require only one debeaking because they are sent to market before their beaks grow back. Most egg producers debeak their

A chick being debeaked.

Debeaked hens.

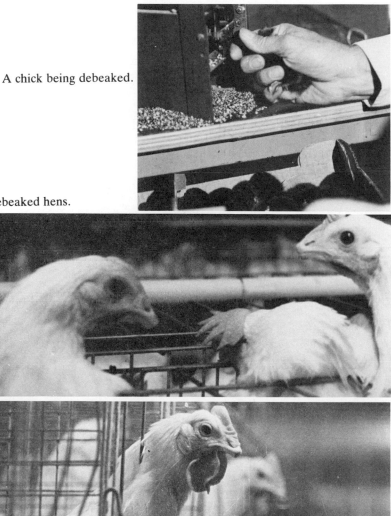

USDA photo

> *Sometimes the irregular growth of beaks on debeaked birds makes it difficult or impossible to drink where a normal bird would have no trouble.*
> —C. I. Angstrom, "Mechanical
> Failures Plague Cage-Layers,"
> *Onondaga County Farm News*
> (Syracuse, N.Y.), December 1970, p. 13

birds twice, once at about one week of age and again during the growing period when the birds are between twelve and twenty weeks of age.

In view of the size of flocks and the cost of labor, the debeaking procedure must be carried out as quickly as possible, with experts recommending a speed of about fifteen birds a minute. But patience and precision tend to give way in monotonous work, and the beaks of many birds are sloppily cut.

> An excessively hot blade causes blisters in the mouth. A cold and or dull blade may cause the development of a fleshy, bulb-like growth on the end of the mandible. Such growths are very sensitive and will cause below average performance. . . . Incomplete severance causes torn tissue in the roof of the mouth. The bird's tongue must be held away from the blade. Burned or severed tongues result in cull [worthless] hens.[6]

Even if debeaking is "properly" done, it is painful and can affect birds' health later. Some debeaked birds do poorly during the production cycle and do not grow to full size because "beak tenderness" makes it difficult for them to eat and drink.[7] At the same time the birds are debeaked, on some farms their toes are clipped just behind the claw using the same hot-knife machine. This operation is said to keep the birds quieter as it prevents "back-ripping" and fighting. To hold down pecking and fighting among males on breeding farms, producers usually cut off their wattles and combs. As these "dubbed" males are not able to easily recognize each other, there is less competition for social position.

Getting Rid of Bugs

The battle against bacteria calls for strict measures throughout the factory. Everyone—animals, managers, and visitors—must follow a one-way route from building to building to avoid bringing germs back to younger animals. Between "crops" of animals, farmers sterilize practically everything inside with an arsenal of hot water, high-pressure hoses, acids, cleansers, and disinfectant chemicals. Animal disease experts recommend "health programs"—routine doses of sulfa, antibiotics, vitamins, and other medication at regular intervals throughout the production cycle—to help hold down disease losses. In addition to routine medication, factory animals receive other doses of antibiotics, drugs, and medicines when specific health problems occur. The mass-production schedule does not allow for precise, individualized treatment and so many producers use a "shotgun" approach to symptoms of disease. According to one veterinarian: "Often the wrong antibiotic is used, a high enough level is not used, or the correct level is not used long enough."[8]

Arrival:
  1) physical exam of each calf
  2) rectal temperature of each calf
  3) nasal and fecal swabs of five selected calves
  4) selenium (Mu-se)® ½cc IM once
  5) vitamins A, D, and E (Injacom)® 2 cc IM once
  6) injectable iron (Ferrextran)® 5 cc IM once
  7) injectable B-complex vitamins (Betaplex V)® 3 cc IM per day for 4 days
  8) injectable antibiotics as indicated
  9) bicarb and dextrose in milk as prescribed with antibiotics, if needed, on basis of cultures
  10) close observation of calves 4 to 5 times a day for first 2 weeks
  11) delouse using Korlan II, observe closely

A schedule of drugs and chemicals administered on the first day of a typical factory "health program."

Producers must also use pesticides to get rid of the mites, ticks, fleas, chiggers, and other insects that build up around factories. Some egg producers use power sprayers that roll along the aisles, shooting mists of insecticide up through the cage floors onto the birds. Chemicals added to the feed aid in the control of flies; they pass through the birds' digestive tracts and remain active in the manure, where they kill fly larvae.

Keeping the Biomachines at Work

At the core of factory farming are those controls or manipulations that are intended to push up production while holding down costs. The business of animal farming is being brought squarely into the later twentieth century by experts who scorn traditional methods as they explore every scientific nook and technological cranny for ways to "improve" it. With neither exaggeration nor political bias, it is accurate to say that "improvement" is measured solely by profitability. In the literature of factory farming, old-fashioned animal management practices

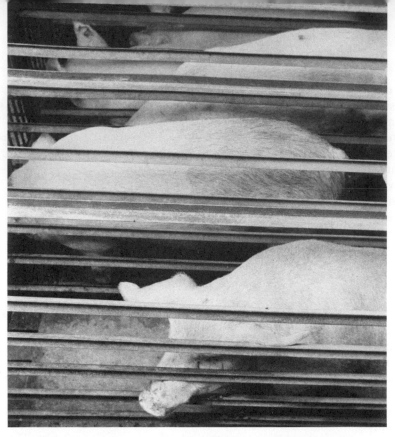

Pregnant sows are confined in stalls for nearly four months.

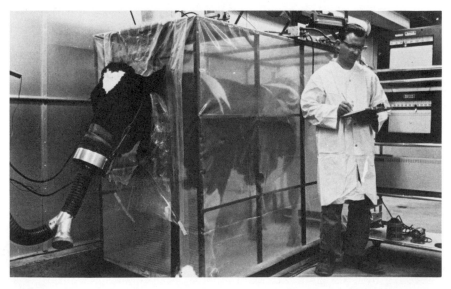

A study on metabolism in dairy cows—part of the quest for high-producing animals.

are deemed to require too much labor; natural animal reproductive and physiological cycles are too slow and unpredictable; and much of animal behavior is just plain unbusinesslike. Animals had it easy on traditional farms; now they must be made to pay more dearly for their brief keep.

Restricting animal mobility is a fundamental tenet of factory management. It is both an unintentional consequence of the economic pressure to crowd factory animals and a deliberate means of manipulating them toward greater productivity. Chickens, if not confined, would doubtless prefer to lay their eggs in nests of their own making rather than on wire-mesh floors crowded with a half dozen other birds. But the costs of nest materials, space, and the labor required to track down and collect eggs from nests would not permit large-scale factory production. Layer hens must be put in cages designed to permit mechanical egg collection. Similarly, narrow stalls confine calves to ensure that flesh of the right color and texture develops rapidly. Farrowing stalls and tethers hold sows in place to allow a maximum number of piglets to suckle and grow to meaty maturity.

Together with other manipulations, restriction of movement contributes to the illusion of greater productivity in factory animals. The "high feed efficiency" boasted of by factory experts is achieved by feeding high-calorie mixtures of ground corn, soy meal, and other nutrients to animals virtually immobilized in warm buildings. During cold weather, pig and broiler producers heat their buildings because growing animals put on weight more quickly if they don't have to burn food energy to keep warm. Nor can factory animals "waste" food energy on exercise in their cramped quarters; all activities must defer to cheap, rapid production of tissue.

Building a Better Biomachine

Genetics is the factory farmer's most effective tool. For thousands of years, farmers have selected the "best" animals of each generation to be the breeders of the next. The criteria for selection once ranged broadly over a great number of factors because earlier animals had multiple uses and, without modern drugs and medicines, their health had to be reliable. In those days, the all-around animal was the best animal. But over the years, especially after market economies arose and animal products became commodities, farmers began to select primarily for productivity.

Through breeding, animals have now been split up into subspecies according to what they produce: we have "beef" cattle and "dairy" cattle, "egg-type" chickens and "meat-type" chickens. Lately, this emphasis on productivity has intensified. In dairy cattle, for example, awards and championships are awarded to farmers whose cows produce the most pounds of milk annually. Layer hens of the 1930s produced an average of 121 eggs a year. Through genetics, today's hens nearly double that figure.

But apparently this is not productive enough, for scientists at the University of Missouri are designing a "superchicken" that will lay an egg a day.[9] Elsewhere, animal scientists use computers to search for more profitable genes and work with sex control, cloning, somatic cell hybridization, gene grafting, and gene transfer in attempts to build souped-up commercial strains of animals.[10]

"Bugs" in the New Models

Emphasis on speed and volume of production has brought unexpected problems in factory animals. When humans attempt to take over the entire process of mate selection (a complex process that has been millions of years in the making) for their own profit, many important traits get ignored and eventually the genetically modified animals can go haywire. High-producing dairy cows are tense, nervous, hyperactive animals. Fleshy bodies of broiler chickens and pigs grow heavy so quickly that development of their bones and joints can't keep up. Skeletal disorders are common. Many of these animals crouch or hobble about in pain on flawed feet and legs. The pig breeder's emphasis on larger litters and heavier bodies, coupled with a lack of attention to reproductive traits, has produced poor mothering traits in sows and high birth mortality in their pigs. These new, improved females produce such large litters that they can't take care of every piglet. To cure this problem, producers began to select sows with a greater number of nipples—only to discover that the extra nipples don't work because there's not enough mammary tissue to go around.

Some of the genetic manipulations for higher productivity seem to work against factory farmers. In beef cattle, for example, the fastest gainers are not the most efficient converters of feed to flesh; slower gainers are more efficient because they eat less and waste less. Although chickens are laying more eggs than ever before, the eggs are different. Factory eggs now are smaller, with more white and less yolk than eggs produced a decade ago; they are also paler and more watery than eggs from barnyard chickens.[11]

Engineering Animals for the Factory

Animal engineers are using genetics to tailor special models for the factory production line. "Minihens" about two thirds the size of ordinary hens enable egg producers to house more birds per cage. Because machines for beheading and defeathering chickens often leave a ring of neck feathers that must be removed by hand, scientists at the University of Georgia tried to develop a "naked-neck" chicken.[12] And to have a truly global industrial chicken, the poultry industry needs a bird that can flourish in hot climates. Therefore, at least six universities in the United

"Silo-type" egg factory, New Mexico.

States and Canada are working to produce a model of commercial chicken with no feathers.[13] Apparently feather research captures the scientific imagination, for there have been attempts to produce a commercial chicken with "bangs"—head feathers long enough to fall over the eyes.[14] These built-in blinders, it was hypothesized, would screen out distractions, leaving serene, calm birds that would gain weight more quickly and would have fewer bruises at the time of slaughter.

Pigs, too, are being modified to fit the factory. Breeding experts are trying to create pigs that have flat rumps, level backs, even toes, and other features that will hold up better under factory conditions. Pigs' natural, physiological reactions to stress are seen by factory experts as abnormalities that must be eliminated; to them, normal pigs are "stress prone." Current efforts to build the better hog include searching for animals that release the lowest levels of ACTH, cortisol, and the other hormones related to the stress reaction.[15] Like its broiler prototype, the stress-resistant pig is the perfect factory animal—a lethargic creature whose only abilities are eating and gaining weight.

The Joy of Sex, Factory Style

Control over genetic traits requires strict control of the bodies and sexual behavior of both male and female animals. Artificial insemination offers producers much greater control over genetic selection and the timing of births; it also eliminates the burden and expense of keeping quality breeding males. The technique is widely used on dairy cattle; it is used less extensively on beef cattle, pigs, and turkeys. It is successful, however, only if the female receives semen at the right time during her estrous cycle. And timing is critical. In pigs, for example, the right moment may last for only a few minutes. If the artificial inseminator misses, the farmer must bear the expense of maintaining the "useless" female until her next heat period and, of course, the "crop" is delayed. Ordinarily the breeding animals themselves determine the right moment by sounds, smells, and other signals. To make artificial insemination more reliable, then, it helps to have a few good, old-fashioned males around to take care of this business—up to a point.

These males are supposed to detect, not copulate with or fertilize females in heat. To ensure that their inferior sperm does not get ahead of the artificial inseminator's vials and tubes, their penises must be neutralized. On some cattle operations, farmers block the penis with a removable plastic tube and a stainless steel pin through the bull's sheath. Because this device holds the penis inside the animal, soreness and infection may set in and the bull may lose his desire to mount. Other producers, then, prefer the permanent solution of penectomy, surgical removal of the penis (these males are called "gomer bulls"). At a midwestern pig factory, boars' penises are surgically rerouted to exit their bodies at the flank. These males are called "sidewinders."[16]

Most factory farmers who use artificial insemination, however, would rather eliminate boars entirely. They consume feed and take up space. After all, the whole boar is unnecessary; all that is needed is his "essence." Not surprisingly, someone has now come up with an aphrodisiac for pigs handily packaged in an aerosol spray can. The product, called "Boar Mate," is marketed by a British firm and contains a synthetic hormone similar to the sex-odor pheromones that boars emanate when they arc in the presence of a sow in estrus. At breeding time, a few spurts from the can of "instant boar" directed around the sow's snout are said to accelerate her heat period and improve the chances of fertilization by artificial insemination.

Super Sex and Reproduction

Research on the biochemistry of animal reproduction has recently begun to get much more attention in animal science because it is turning out new techniques for increasing "crops." This is one of the few remaining new frontiers for factory animal science in its quest for production efficiencies, for little else of animals' life and life cycles remains uncontrolled. Under one new technique, called superovulation, producers inject hormones into females to stimulate their ovaries into producing a large number of eggs instead of the usual one or two. Another technique, called synchronization, uses hormones to synchronize the estrous cycles of two or more females. In combination with others, these techniques are used by animal scientists and producers to make reproduction of high-producing animals cheap and fast. At the University of Missouri, scientists have been trying to unite in test tubes sperm and eggs taken from "super" breeding animals.[17] The newly fertilized eggs would be surgically implanted in ordinary females for gestation.

In another new technique, a pedigreed cow is dosed with hormones to cause superovulation and her eggs are fertilized by artificial insemination. After a few days they (now embryos) are removed, sorted by sex (by microscopic examination of chromosomes), and implanted into the wombs of ordinary cows. This procedure, called ova or embryo transfer, is catching on with producers of pigs and dairy cattle. In cattle, the high-producing cow's eggs are flushed out by sterile water and then implanted in recipient cows through a small incision in the flank. In pigs, under present methods, the eggs are removed through an incision in the donor sow's abdomen. If her eggs are in demand, she may be cut open and sewn up six or more times in a year.

Although the procedures can cause shock or death, hormone control of estrus, ovulation, gestation, and birth give greater control over the entire factory operation. Estrus control decreases time between pregnancies, aids assembly-line artificial inseminations, increases the chances of conception, and makes planning and record keeping easier. Use of prostaglandins to induce labor contractions makes calving and farrowing more convenient for the farmer. Injections of progestins or steroids bring on twin calves, larger litters of pigs, and bigger lamb "crops." As soon as they give birth, females can be dosed again with hormones to tune in their cycles for rapid rebreeding.

"It would not be unusual to perform enough [ova transplant] surgeries on an individual [sow] in a year to equal the number of offspring that she would produce in a lifetime of normal farrowings."—Jason James, D.V.M., of the Sullivan Veterinary Clinic, Sullivan, Ill., quoted in *National Hog Farmer,* December 1975, p. 28.

"It is obvious that the light supplied by sunshine during the day and normal darkness at night is the most inferior of any lighting program."—Mack O. North, "A Case Can Be Made for Continuous Lighting," *Broiler Industry,* September 1976, p. 48.

Speeding Up the Assembly Line

Even without the use of drugs, farmers speed up reproductive cycles by separating calves, lambs, and pigs from their mothers much earlier than nature would. In nature, a calf might nurse and run with its mother for about a year; on a dairy farm, a calf is lucky to spend more than a day with its mother. Although most factory pig farmers leave their sows and pigs together for about three weeks before separation and weaning, a few are trying to wean only a few days after birth in order to rebreed sooner.

In addition to manipulations of sex and reproduction, factory experts control growth rates to increase production. The poultry industry has known for some time that birds' rates of growth and egg laying depend on the daily change in the ratio of light to dark. In the spring, when days grow longer and nights become shorter, birds' body cycles pick up and their rates of egg laying increase. It didn't take poultry producers long to figure out that control over light meant control over production. They began to experiment with various light schedules. Some broiler producers have total control over light in their windowless houses; others take advantage of sunlight during the day and use artificial lights after dark. Egg producers try to create the illusion of eternal spring by keeping the lights on a little longer each day. After about a year of this, the flock's productivity drops, and many producers use "force molting" to revive it. This technique shocks and disorients the birds by leaving them in the dark for a few days without food or water. A few birds die in the process, but most come through and begin producing all over again on a renewed pseudo-spring light routine.

Other attempts to boost productivity involve the substances fed to animals. Not the least of these is the high-calorie feed itself; but feed additives are the mainstay of factory productivity. Antibiotics are the most widely used. Each year, about 9 million pounds of antibiotics go into animal feeds.[18] Exactly how antibiotics help animals put on pounds is not known, but some experts believe they provide a shortcut around good animal care. According to one expert, "the 'growth promoting' effect of antibiotics is seen only when animals are raised under suboptimal conditions—that is, in crowded, dirty and heavily contaminated pens and feedlots."[19] To get rapid gains in feedlot cattle, many operators also implant in their animals' ears pellets containing synthetic hormones and other growth-promoting compounds. The most commonly used implants are Synovex, Ralgro, and DES (diethylstilbestrol)—see page 69.

Cheaper Raw Materials

Every so often new gadgets and ideas come out that are claimed to boost production. There is a nylon device with extending prongs like an umbrella that is implanted in the vaginas of young feedlot cows. The device is believed to enhance growth by stimulating nerve endings in the vaginal tract, creating biochemical changes like those of pregnancy. But no one really knows for sure how it works; all they know is that it produces an average increase in profits of $12 to $14 per head—and more if used with chemical growth stimulants. Scientists at the University of Arizona are studying the body chemistry that curbs appetite when a beef animal eats its fill: "Obviously, if the thing that turns a beef animal away from the feed bunk were found and could be overcome, it would mean a lot. . . . It could even mean an added supply of beef at a cheaper price." [20] Even ordinary cement dust may soon be an additive in cattle feed because, according to the U.S. Department of Agriculture, it produces weight gains 30 percent faster than cattle on regular feed.

Controls over appetite and feed consumption are much sought after now that animal scientists see their relationship to productivity. Researchers in Europe and Australia are breeding for animals that eat, chew, and digest faster, as these traits seem to indicate high productivity. These experimenters are also depriving cattle of sleep to make them hungrier, heavier eaters that will produce more meat and milk, and they are destroying sheeps' sense of smell to make them heartier, less-finicky eaters of cheap commercial feeds. [21]

In their efforts to cut costs, factory experts try almost anything cheap and disposable as feed for animals. Ground-up cardboard is reportedly being promoted by a Chicago firm as "an exciting development" in animal feeds. "The cost savings would be tremendous," the firm was reported as claiming, because some 10 million tons of cardboard would be saved from the dump and channeled through dairy cattle into food for humans. [22] Old newspapers have also been suggested for recycling through cattle. A Missouri farmer fattens his cattle on a diet of 75 percent sawdust laced with ammonia.

Recycling of animal wastes back into feed is another trend in dietary cost cutting. In about a dozen states, mostly in the broiler country of the Southeast, recycled animal wastes have been approved for shipment and sale intrastate as a feed material. In other states, producers can recycle waste on their own farms without violating any laws. One factory expert admits that the motivation behind waste recycling is "raw and simple economics." [23] It cuts feed bills and reduces disposal problems. A few producers' systems for waste recycling are simple and direct: one Pennsylvania farmer runs pigs in the pit under his caged-layer factory. A Kansas farmer runs about a hundred pregnant sows in the waste pits under his finishing pens and boasts that, by not having to feed them for ninety days, he saves about $9,300 each year on feed. [24] On some factory farms, dirty litter from broiler houses is scraped up, hauled away, and added directly to cattle feed. Raw poultry and pig manure is mixed with ground corn or shredded stalks and fed to pigs and cattle on some farms. Animal waste is also processed and sold as a cheap feed supplement by agribusiness companies. Elsewhere, animal scientists are trying to perfect the "oxidation ditch," which channels the liquid wastes from factory manure pits back to the animals; they have to drink it because it's the only "water" offered to them.[25] Even human sewage is being studied for ways to process it into animal feed.

Spreading the Cost of the Factory

Economic considerations also lead to severe crowding of animals in factories. Animal scientists have shown that the animal death rate increases as crowding increases but that these losses are insignificant compared to the much greater yield of meat or eggs per unit of space, labor, and overhead. In the earliest days of the broiler industry, when birds began to be raised indoors, "Scientists went to work to determine the optimum amount of floor space needed per bird."[26] The study of the economics of crowding had begun.

Dozens of experiments were reported over the next several years. They showed that reducing the floor space below one square foot per bird also reduced growth and feed conversion. But conflicting evidence showed that more pounds of chicken could be produced in a house when the birds were "crowded."[27]

The table below shows how "more pounds of live broiler can be produced per square foot if floor space per bird is reduced."[28]

Effect of Floor Space on Weight, Mortality, and Pounds of Birds Produced per Square Foot

Floor space per bird (square feet)	Average live weight (pounds)	Mortality (percent)	Weight of birds raised per square foot of floor space
1.0	4.12	2.1	4.12
.9	4.09	2.3	4.53
.8	4.05	2.6	5.03
.7	4.00	3.0	5.66
.6	3.94	3.6	6.40
.5	3.86	4.5	7.54
.4	3.75	5.8	9.03

Source: Mack O. North, "Some Tips on Floor Space and Profits," *Broiler Industry,* December 1975, p. 24.

"There's too much wasted space in a typical controlled-environment single-deck nursery. The cost of the building is just too big a cost factor. Stacking the decks spreads that building cost out over more pigs."—John Byrnes, "Stacking 3 Decks of Pigs," *Hog Farm Management*, January 1978, p. 16.

We saw many units being built at a cost of $1,000 per sow capacity. You've got to get as many pigs as possible out of a sow in quarters like these.
—G. R. Carlisle, "Who Is Going to Produce the Seedstock?" *Farm Journal*, October 1976, p. Hog-10

Economics determines the limits of crowding in other factories as well. When two Iowa cattle producers tried to put eighty animals per pen in their new confinement building, the cattle started having arthritis and joint problems. The producers realized that the animals weren't getting enough sunlight and exercise, so they reduced the number per pen from eighty to seventy and started feeding mineral supplements. According to one producer, "After we did this, we stopped our leg and feet problems."[29] But this change increased the investment cost per animal by $17.86 and lowered the building's overall capacity and profitability.[30]

The trick is to find the maximum number of animals that can be crowded into the factory without stress and disease taking such a high toll that profits go down. Up to that point, stress and disease do not matter. One study on the "Effects of Density on Caged Layers" showed that a higher degree of crowding in cages, even though it pushed up mortality, produced better profits if the price of eggs was above a given point. According to the authors, "When the farm price of all eggs is 40¢ or more . . . five layers per [12-by-18-inch] cage make a greater profit."[31] In other words, when eggs are dear, hens are cheap.

This simple economic fact gives the lie to claims that since animals in factory farms are productive, they must live happy, contented lives. Even if it were true that putting on weight is a sign of well-being—which it is not—productivity of individual animals is not what counts in factory farm economics. As the studies above illustrate, individual animals' well-being and productivity can suffer so long as more pounds of meat or dozens of eggs can be squeezed from a factory building.

Factory experts show no signs of giving up their search for further tinkering to solve problems in the factory. Too much has been invested to think of turning back. Yet animal efficiency in commodity production has biological limits. Animal scientists diligently pursue their work of modifying animals, systems, and methods, turning out round after round of "improvements." But the excitement over early gains evaporates as it becomes obvious that those gains are diminishing. Factory farming has not brought the overall efficiency claimed by its proponents.

- Increases in milk production per cow have flattened out since 1972.
- Egg production per hen is leveling off at about 230 per year.
- Pigs saved per sow have actually decreased since 1969.[32]

Factory farming managers have not yet learned the lesson that Rachel Carson tried to teach those who would attempt to manage the natural environment: biological mechanisms are more complicated than we realize, and our attempts to manipulate them in our own interests are likely to have unexpected costs.

THE FACTORY

Are Biomachines Good to Eat?

Federal efforts to protect consumers from illegal and potentially harmful residues of animal drugs, pesticides, and environmental contaminants in raw meat and poultry have not been effective. GAO estimates that 14 percent by dressed weight of the meat and poultry sampled by the Department of Agriculture between 1974 and 1976 contained illegal residues.
—Problems in Preventing the Marketing
of Raw Meat and Poultry Containing
Potentially Harmful Residues
(Washington, D.C.: Comptroller General
of the United States, April 17, 1979), p. i

We've been accused of selling a chicken with less flavor than the "old-time" chicken. . . . Attempts are being made at overcoming the flavor problem by injection or marinating.
—W. J. Stadelman, "Old-Time Flavor:
New Injectables Possible," *Broiler
Industry,* April 1975, p. 79

Since the main end product of animal factories is food for humans, we should take a close look at the relationship between current farming methods and food quality. Reflect for a moment on the farms of yesterday—when rates of animal growth and reproduction were determined by nature rather than by mortgages, and animal health came from sunlight and exercise rather than from injections and additives. Typical farms kept a few milk cows to provide fresh milk and butter for the family and sold the small surplus to markets nearby. Nearly every farm had a flock of chickens that roamed barnyards, fields, and pens looking for seeds, worms, and insects. Freedom of movement was allowed not out of any humanitarian motives but simply because farmers would not waste grain on scavenger chickens. A few hogs nourished on table scraps and food rooted from the earth could provide the family with lard, sausage, hams, and bacon. These animals, though valuable to the farm family,

53

were not regarded merely as cash on the hoof; because they were comparatively few in number, the health and well-being of individual animals was important to the family's survival and not just a matter of profits. These animals were "domestic" in the precise sense: they were part of the rural household.

Though closely watched, yesterday's farm animals were not completely under the farmer's control. Many, of course, were in pens or enclosures and some were subject to abuse and neglect. But for the most part they had space, daylight, seasons, social interaction, free range about the fields for choice of forage, resting areas—in other words, some degree of what humans might call freedom. The farmer expected no more from them than they could naturally produce if provided with food, water, and shelter.

Freedom for animals within their environment had advantages for people, too. The health of animals was not so precarious and expensive to maintain as it is now. Contact with the soil gradually exposed animals to a range of microorganisms, allowing their natural immunity systems to work and combat disease effectively. Roving animals ate a variety of foods as their appetites dictated, and this diversity in nutrients and exercise produced a quality in foods not found today. (Compare the taste, appearance, and nutritional content of an egg from a "free-range" hen with one from the supermarket.) And because they sustained themselves on insects and plants inedible to humans, free-ranging animals did not compete with humans for food to the degree that most farm animals do today. Unconfined and dispersed over the land, these animals distributed their wastes to enrich the soil without pollution.

Fast Food

Today's animal-rearing methods are geared to mass production. Their environment now consists of cages, steel bars, fluorescent lights, dusty air, and total darkness except at feeding times. Animals' health and vitality, mental and physical, suffer. The quality of the flesh, milk, and eggs they produce also suffers. Chemically dosed and genetically hyped up to meet factory standards of profitability, today's animals are artificial to the marrow.

The poultry industry, the trend setter in factory ways, has known for some time that the factory's impact on animals depresses food quality. It uses chemistry to impart flavor, color, and other characteristics lost or distorted in the factory process. The food we get from birds reared indoors tends to be pale and washed out compared with that from birds raised in natural daylight. Most commercial poultry operations therefore use feed additives containing xanthophyll or beta-carotene to enhance the yellow color of skin and egg yolks. The brighter color may be attractive to

An egg assembly
line—three levels
high.

the shopper's eye, but there is evidence that color additives actually lower
the quality of eggs.[1] More important, eggs from caged layers may be lower
in some nutrients—particularly vitamin B_{12} and folic acid—than eggs from
free-ranging hens.[2]

In other animals, the factory's emphasis on fast weight gain from high-
energy feeds lowers nutritional quality by increasing production of
storage-type (concentrated, saturated) fats but not of protein, "structural" fat, and other nutritious material. A grain-fed feedlot steer, for
example, will produce a carcass that is about 30 percent storage-type fat
and 50 percent "lean." The "lean" portion contains from 7 to 20 percent
of its fresh weight as saturated, storage-type, triglyceride fats—the
"marbling" in your steak. Unlike structural fat, which is highly unsatur-
ated and contains essential fatty acids, this kind of fat is waistline fat—
nutritionally nonessential and a contributor to various health problems for
its consumer. At best, then, only about 45 percent of the feedlot animal's
carcass is actual muscle cell and if muscle water is excluded, only about 9
percent of the entire carcass is nutritious material. In the factory-farmed
steer, there is about three times as much nonessential, storage-type fat as
there is nutrient material; by way of contrast, in free-living animals there

is between five and ten times as much nutrient material as there is storage fat.[3] According to an expert on nutrition:

> The extreme end product to modern intensive husbandry, "the heavy hog" is a method of fat production; this system seems pointless in terms of satisfying nutrient requirements of the human population the animal industry is meant to serve.[4]

Chemical Feast

For some time now, the turkey industry has used injections containing flavorings and phosphates to improve the taste and texture of turkey meat.[5] Manipulation of chicken flavor through chemistry may soon be used by the biggest producers to establish a "unique identity" for their products.[6] Says one expert, "It should be possible to uncover a material or materials that could impart that 'old barnyard' flavor in chickens."[7]

Like most of the other food industries, animal products industries rely heavily on chemistry from soil to supermarket. Some of the uses of chemistry are:

*AS COLORING AGENTS Xanthophyll, zeaxanthin, marigold petals or extracts of them, and caroteneic acid compounds are some of the feed additives that make chicken skins and egg yolks yellow.

*AS ANTISPOILANTS   Vitamin E is sometimes added to turkey and broiler feed to delay spoilage in the carcass after slaughter and storage. Fungicides and insecticides like malathion are additives used to retard spoilage in stored grain and feed.

*AS FLAVORING AGENTS   Various additives such as Hog-Krave are added to feed to stimulate appetites in factory animals when feed consumption falls off during hot weather or other times of stress. Chemical appetite control works the other way, too. Compounds to inhibit appetites are used to cut feed costs on animals, such as gestating sows, that are not being fattened for market.

*TO CONTROL PESTS   Producers must routinely apply pesticides to control flies, fleas, ticks, mites, and other pests that build up around the concentrations of animals, manure, and grain on factory farms. Besides the usual ones, a "new wave" of chemicals is on the way to "really knock out pests which have developed resistance to some standard insecticides," according to one farming magazine.[8] These include new families of chemicals, such as the synthetic pyrethroids, new growth inhibitors, and new variations of old-school hydrocarbons and organophosphates.

*TO CONTROL DISEASE   More than 40 percent of the antibiotics and other antibacterials produced annually in the United States are

Next in line at a large factory dairy: milk is a natural?

used as animal feed additives and for other nonhuman purposes. Nearly 100 percent of poultry, 90 percent of pigs and veal calves, and 60 percent of cattle get antibacterial additives in their feed.[9] Seventy-five percent of the hogs that are marketed eat feed laced with sulfa drugs.[10] Most factory dairies use one of about sixty kinds of chemical teat dips after each milking to reduce the spread of mastitis (udder inflammation) in their herds. There is evidence that some of these dips leave residues in the milk that may be dangerous to humans.[11] The Iodophor teat dips most commonly used in the United States have been banned in Germany because they leave iodine residues in milk. In most poultry processing plants, freshly killed carcasses are immersed in "wash water" laced with chlorine to kill salmonella bacteria. The European Common Market countries have banned this type of wash because they say it only contaminates carcasses.

　*TO BOOST PRODUCTION Antibiotics and hormones such as DES† are the basic chemicals in the factory farmer's arsenal of

---

†On November 1, 1979, the use of DES in food animals became illegal. See text, page 69.

productivity boosters, but there are many others. Shell Oil Company manufactures a new feed additive called XLP-30 that is supposed to "boost pigs per litter." "We don't know 'why' it works," says a company official, but Shell advises that for every dollar spent on the chemical $7 to $11 in returns can be expected.[12] Many dairy producers use "infusion programs"—administrations of antibiotics three to four weeks before breeding—to boost chances of fertilization. Beef experts are trying out probiotics (to aid growth of favorable bacteria in animals' digestive tracts), estradiol (a hormone), monensin sodium, and for-maldehyde (a disinfectant and preservative) in feed to speed up weight gains. About 70 percent of U.S. beef by carcass weight comes from cattle that have been fed growth-promoting feed additives.[13] Antioxi-dants—chemicals that inhibit the formation of peroxides during metab-olism—are added to feed in many poultry operations to lengthen layers' production cycles. Antioxidants are also used to increase the effectiveness of other additives such as coloring agents and vitamins. Electrolytes—common salt and other chemicals that increase the electrical conductivity of water solutions—are used to pep up stressed birds and stimulate appetites. They are used also to increase effective-ness of other additives such as antibiotics, sulfas, and nitrofurans (antibacterials). Small amounts of arsenicals, in the form of arsonic or arsanilic acid, are used by most poultry producers to speed maturation, increase feed efficiency, and stimulate egg production. Chemical capons are now made by implanting chickens with the hormone estradiol which, according to a supplier, "causes them to have a docile and peaceable disposition" and turns energy to weight gain.[14]

In all, there are over 1,000 drug products and as many chemicals approved by the Food and Drug Administration now in use by livestock and poultry producers.[15] Many leave dangerous residues in animal products if not withdrawn from the animals before slaughter or before milk is collected.

Changing the "art" of growing chickens from a backyard enterprise to a "science" has involved chemistry in many ways.
—Milton L. Sunde, "The 'Chemical Feast' That Helped Us Grow," *Broiler Industry,* March 1977, p. 54

Deceiving Consumers

Reliance on chemistry, along with other factory practices, deceives consumers into thinking animal products have better quality than they really have. What other purpose would skin and yolk yellowers serve? Producers try to dodge the issue by saying that the purpose is to enhance "eye appeal," which is only another way of saying that their products sell better if they have the look that consumers believe to be associated with quality. To end this deception, the Food and Drug Administration proposes to require producers to state on their labels whether colorings are used.

Purveyors for animal factories know what chords to strike to move their products. In spite of the extensive use of chemistry in egg production, one poultry industry leader advises his colleagues: "Slant egg carton copy along this line. 'Eggs are a health food. A natural human food. No additives, no preservatives.'"[16]

As we saw in the last chapter, factory genetics has diluted egg quality over the years. When manufacturers of flu vaccine needed high-quality eggs back in 1976, they didn't buy eggs from factory farms but from Amish farms with smaller flocks and manual labor methods. Big broiler companies are rich enough to afford slick advertising campaigns that allege the superiority of their brand of chicken, but according to Consumers Union, the allegations don't hold up. In recent tests, a Consumers Union panel found that the heavily advertised brand-name broilers were no plumper or better tasting than the unbranded super-market ones, although they were priced 9 to 14 percent higher per pound.[17]

Our canned hams are fattier and flabbier than European brands, and an American meat-packer believes it is "due to corn-feeding and controlled-environment housing in the U.S."[18] The flesh of these cramped, corn-fed pigs turns to fat and the bacon they produce nearly disappears in the skillet. When slaughtered, stressed pigs yield flesh that inspectors and packers call PSE (*p*ale, *s*oft, *e*xudative) because of its appearance and the watery ooze it gives off. More and more beef these days ends up as hamburger probably because that's the only way industry can get rid of the fatty meat produced by feedlot methods.

Let Them Eat Carcinogens

Factory methods do worse than rob consumers of quality: they also expose them to greater risks than ever before. Mechanization, chemistry, and greater centralization may reduce the costs of feeding animals, but it multiplies the risk of harm from what might otherwise be a simple accident. In June 1973 several hundred pounds of a flame retardant

Burying mistakes: disposal of poisoned cattle after the 1973 PBB incident in Michigan.

Dan Perszyk/NYT Pictures

containing cancer-causing polybrominated biphenyls (PBBs) were accidentally dumped into animal feed at one of the Michigan Farm Bureau's feed mills. As a result, more than 30,000 cattle, 2 million chickens, and thousands of sheep and pigs died or had to be killed by farmers. Through bureaucratic inertia or incompetence, state authorities did not withdraw the feed or trace its source until ten months later when the poison was identified by a U.S. Department of Agriculture chemist. By that time, the PBB-poisoned meat, milk, and eggs had gone out to consumers. By the summer of 1976, 96 percent of nursing mothers tested in Michigan had PBB in their milk. PBB in the manure of contaminated animals ended up in the soil, lakes, and rivers. Michigan vegetables began to show residues of the carcinogenic chemical. According to testimony before Michigan's Senate Commerce Subcommittee hearings on March 29, 1977, nearly all Michigan residents now have intolerable levels of PBB.[19]

An isolated incident? One authority on toxic substances believes that "PBB represents the kind of problem we're beginning to see not only in Michigan but in the United States and many parts of the world." [20]

Routine use of drugs and chemicals in the factory process is the major contributor to dangerous residues in animal products. Before November 1979, DES (diethylstilbestrol) was still being used by the cattle industry, although there was plenty of long-standing evidence that it is carcinogenic in humans. Dangerous residues of drugs, pesticides, and chemicals are showing up more and more in animal products, and government measures to protect us are not effective. According to a recent report by the General Accounting Office, 14 percent of all meat and poultry sampled by the Department of Agriculture between 1974 and 1976 contained illegally high levels of drugs and pesticides. The report notes that many of the substances used around food animals are known to be dangerous.

> Of the 143 drugs and pesticides GAO identified as likely to leave residues in raw meat and poultry, 42 are known to cause or are suspected of causing cancer; 20 of causing birth defects; and 6 of causing mutations. [21]

Antibiotic residues in milk are causing allergic reactions in some people because of routine teat-dipping and infusion programs in modern dairy factories. Chlorinated hydrocarbons used on corn, soybeans, and other grains accumulate in the fat of broilers and can remain at high levels for five weeks after the birds are put on clean feed.

The factory's search for cheaper feedstuffs may cause additional health problems for consumers. According to a veterinarian who became a meat inspector after twenty years of farm animal practice, "There are a number of major human diseases, namely, cancer, heart disease, and gallstones, that . . . originate in the meat-packing plants of this country." He believes that animal feed manufacturing practices contribute to a high incidence of cancer in factory animals because cancerous tissue is recycled.

> What happens to the 15 million pounds of animal tissues which are too severely affected with cancer to be used? They are processed into hog and chicken feed. The result is a recycling of potential cancer substances repeatedly through the human and animal food chain. [22]

Farm industry and pro-farming government officials downplay these risks, arguing that drugs and chemicals are rigorously tested before they go into commercial use. The animal "health products" (read: drug and chemical) industry points to the government's long testing procedures through which new products must go before they can be marketed. But these procedures seem more and more inadequate as safeguards against

I am one of the many practicing veterinarians who witness misuse or misapplication of millions of dollars of drugs on the farm, on a day-to-day basis.

—Letter from F. B. Lederman,
D.V.M., Blue Earth, Minn., in
"Reader Speak Up," *National Hog
Farmer,* November 1976, p. 42

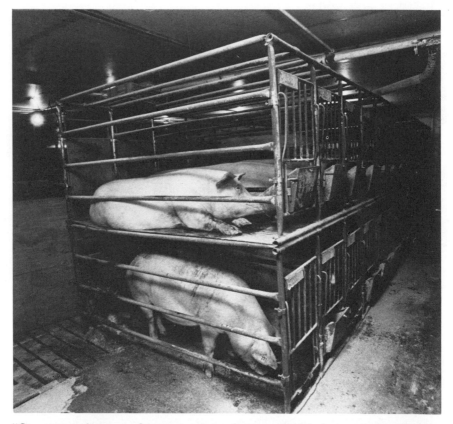

"One reason large confinement systems have worked is because of antibiotics. Without antibiotics it would be hard to have these larger systems and crowd the pigs as we do in some cases."—John Armes, McQuady, Ky., pig producer quoted in "Is Absolute Safety Impossible?" *Hog Farm Management,* March 1978, p. 98.

the kinds of risks posed by today's agrochemistry. DES, DDT, and some of our other drugs and pesticides did not reveal their dangers until years after they had been approved as safe by the government.

Drug Abuse

Farmers feeding drugs to animals are supposed to "withdraw" the drugs (that is, stop feeding them) a specified number of days before the animal is sent for slaughter. The idea is that the drugs should not be present in the animal's flesh by the time it is killed; but farmers aren't always careful enough. Some don't follow directions; some don't stop feeding drugs before sending their animals to slaughter; and then, of course, there are mistakes. Mistakes are much more likely in a factory where various types of feed and additives are used. According to one expert, "It's incredible how many people don't know what additive is in their feed." [23] They aren't helped by the feed and drug manufacturers, who don't always state on their labels what drugs and additives are mixed in. Withdrawal schedules are supposed to help farmers decide when to remove additives before sending the animals to market, but the schedules can be confusing and hard to follow.

As well as the farmers' failure, or inability, to properly withdraw drugs before shipment, the factory machinery appears to contribute to residues in animal products. Large feed-storage bins and long, mechanized tubes and troughs hold quantities of drug-laced feed for several days after the drugs and additives have been withdrawn. According to an FDA study, drug carryover in feeders was responsible for 57 percent of illegally high residues in pigs during the last half of 1977. [24]

Bureaucratic inertia and industry lobbying efforts prevent government protective procedures from keeping up with new problems in factory methods. Waste recycling, for example, is catching on fast as factory farmers search for cheaper feedstuffs and solutions to their monumental waste disposal problems. But because the wastes contain drug residues, the practice adds to the factory farm's problems with its products. According to one expert, "A producer refeeding waste almost has to assume those sows are getting some drugs." [25] Recycled waste also contains levels of toxic, heavy metals such as arsenic, lead, and copper. But recycled waste is not as uniform in content as a commercial drug; its elements vary widely from farm to farm and from species to species. This means that a withdrawal period or dosage in feed that is safe for one batch might not be safe for one from another farm or from other animals. Another factory trend, that toward "flush-floor" systems for waste removal, can cause residue problems when that flush water is recycled. Water-soluble sulfa drugs can build up in the water and be taken in by pigs as they drink from the floors.

Placating Consumers

In 1978, when one eighth of milk and slaughtered pigs around the country showed illegally high levels of sulfa drugs, the animal drug industry and the Department of Agriculture got worried enough to look into the causes and sent out warnings to producers to be more careful with additives. The industry feared that a consumer backlash to this epidemic of residues could lead to regulations on the sale and use of its free-flowing drugs. Their efforts did bring down the number of residue violations somewhat, but the problem will not be eliminated by voluntary programs. Some drug controls must come eventually as the spread of factory farming adds to the incidence of residues in animal products. The farm industries will be forced to either accept some sort of controls or face losses at the supermarkets as consumers look for safer foods.

When these regulations do come, they may do no more than alleviate public anxiety. Regulations on the antibiotics given to farm animals were imposed in Great Britain in 1971 but, according to a British scientist, they "have had little effect on any of the situations they were designed to change."[26] By 1976, British farmers were feeding regulated antibiotics at about the same levels as they were before regulation. The problem may even have worsened since regulation was imposed. Apparently, British farmers are "trying to squeeze every bit of efficiency out of their feeds" by feeding "the few drugs available in massive amounts and in combinations U.S. producers would not even consider."[27]

Black markets on desirable drugs have sprung up as a result of the British regulations also. But then we already have black-marketing of desirable drugs in the United States. According to a farming magazine, "Illegal use of animal prescription drugs has become an alarmingly widespread practice."[28] In 1976, dimetridazole, a drug approved only for turkeys and a suspected carcinogen in humans, was widely but illegally used by midwestern pig farmers to control dysentery.[29]

Old-Time Protection vs. Brave New Farm

The Department of Agriculture's meat and poultry inspection program is inadequate protection against the hazards brought on by today's disease-ridden, drug-dependent factory methods. The basic concept of individual carcass inspections was set down back in 1906 when drug residues were unheard of and the flow of flesh through slaughterhouses was a slow trickle compared to today's torrent. The carcasses of about 120 million cattle, calves, sheep, lambs, and pigs and over 3 billion chickens, turkeys, and ducks now go through the federal inspection lines each year. One departmental inspection service official admits that "USDA has a huge problem in trying to assure wholesomeness of product and still

From the crates to the conveyor line.

Into the killing room.

Bleeding.

Scalded and plucked.

Brian
Baker

accommodate industry's needs to automate processing of 10 billion pounds of production annually."[30] Since the rise of factory farming and mechanization of meat and poultry processing, the department has been struggling to maintain some semblance of adequate inspection under the old law's requirements; there have been revisions since, but the basic method of post mortem inspection has gone unchanged for twenty years.[31]

An inspector in a typical broiler-processing plant has about three seconds to inspect each carcass for twenty-odd diseases transmissible to humans. Because the cost would be prohibitive, the USDA does not inspect each carcass for residues but takes random samples and turns them over to commercial laboratories for analysis. Catching a residue-contaminated carcass by this means is a matter of chance. And by the time the laboratory test results are done, the meat is out in the stores. According to the recent GAO study cited at the beginning of this chapter, the actual incidence of illegal residues in meats and poultry is far greater than that reported by the Agriculture Department. Departmental procedures contribute to its "magic" violation reports: each animal sampled is tested for only one drug or class of drugs rather than for all of the substances supposed to be monitored. Moreover, the USDA does not even test for some 97 drugs and pesticides likely to leave residues—24 of which are known or suspected carcinogens and 17 of which are suspected of causing birth defects.[32]

Not the least of the problems with the present inspection setup is actual or potential corruption of inspectors by the packers and processors whose products they are supposed to inspect. The job of inspecting carcasses is grueling and the hours are long: twelve-hour workdays are common and they may be divided among several plants in an area. To keep the lines rolling, plant operators can influence inspectors to use their considerable discretion. For instance, under the law an inspector's overtime must be paid by the plant. Since plant operators decide how long their plants will operate, they can grant or deny overtime. It has been reported that "reasonable" inspectors are often allowed to take home free meat and are offered liquor and other gifts. Plant operators and inspectors have been prosecuted and convicted for this kind of low-level bribery, but critics say that is is so endemic to the inspection process that it is practically impossible to prevent.[33] Checking up on the inspection process can be difficult for government officials, as the case of Robert Angelotti illustrates. Angelotti, former head of USDA's Food Safety and Quality Service, was forced to resign his post ostensibly because he was accused of claiming reimbursement for two meals that were paid for by others. The underlying reason for getting rid of him, however, appears to be that he went around to meat-packing plants with an expert without notifying plant personnel and inspectors.[34]

Profits First, Caution Maybe

Although the new farming methods present greater risks than ever before, the controllers of these methods can exert more powerful financial and political influence. The ownership of animal products industries is becoming more highly centralized. You can't judge the quality of your eggs, milk, and meat by a visit to Farmer Jones's farm—you have to take the word (in the form of advertising) of an agribusiness corporation or a well-heeled national promotional outfit like the National Pork Producers' Council or the National Dairy Council. These industries already know that the less said about some of their methods, the better. A study paid for by the Animal Health Institute, the "trade association" of sixty manufacturers of factory drugs, chemicals, and feed additives, disclosed that

> consumers are totally unfamiliar with the practices of meat production. Moreover, they tend to resist acquiring such knowledge, and consider meat favorable as a meal/diet essential, but are discouraged from buying and serving foods with which anything they consider unpleasant has been associated.[35]

Throughout factory farm publications, from magazines for farmers to the slick annual reports of Pfizer, Dow Chemical, and the Animal Health Institute, there is a continuous tirade against "consumerism" and government regulation. Farming magazines that depend heavily on advertising placed by drug and chemical manufacturers feverishly attack attempts to control agrochemistry. At least one such magazine, *National Hog Farmer,* acknowledges its self-interest on the question of drug and chemical regulations, admitting that "we obviously have a stake in this matter."[36] Other magazines are not quite so honest and run articles claiming that the loss of antibiotics and other drugs "would cost the swine industry alone over half a billion dollars" and would increase costs to consumers by "at least $1 billion a year according to the most conservative estimates."[37] Their editors attempt to persuade farmers to fight the drug companies' battles; editorials urge readers to "sound off and sound off fast" to government representatives.[38]

Would farmers lose anything if drugs and chemistry went under tighter controls? On the contrary, they would be better off. One expert advises farmers to "cut out the vitamins and the antibiotics" and turn pigs outdoors on pasture when hog prices slide.[39] A study requested by the U.S. Senate's Agriculture Committee shows that farmers would benefit from a ban on antibiotics, tetracyclines, and sulfa drugs. "Total net revenue to farmers would be initially enhanced," it states, encouraging farm expansion for a few years after which production and prices would level off.[40] The report reveals what is perhaps the real reason why agromanufacturers are hostile to a drug ban. "Increased risks associated with feeding livestock in confinement production systems without low-level use of drugs," states the report, "could make such confinement production less viable."[41]

The controversy over the use of DES as a growth promotant set the pattern for attempts to regulate drug abuse: rising public concern about risks of drug use, followed by government action to ban or regulate those uses, followed by industry battles to hold the line against regulation, followed by government inaction. In 1970, DES was linked to vaginal cancer in women, and uses of the drug in humans were restricted. Then, in 1973, after traces of DES were found in the organs of slaughtered animals, the FDA banned the use of the drug in food animals. This ban was reversed by the courts after several drug companies sued the government, saying that the FDA did not follow the proper procedures in imposing the ban. Slowly but surely, the FDA came back with another attempt to ban the drug. This time the agency successfully steered through the administrative thicket and its prohibition on oral and implant uses of DES in food animals became effective on November 1, 1979. But industry, of course, wouldn't take this lying down. As of this writing, two manufacturers of DES have lawsuits pending in the federal courts challenging FDA's conclusions about the dangers of the drug and asking that its ban be lifted.

So the battle goes on, for DES and for the other suspect drugs and chemicals. No doubt it will continue for many years. The eventual outcome is unpredictable. Meanwhile, consumers worried about the quality and safety of factory-farm products have at least one way out: they can quit eating them.

We feed medicated feed all the way to market.
—"Stick with Hogs," *Farm Journal,*
March 1976, p. Hog-17

FACTORY IN REVERSE

Wastage in Factory Production

Expensive facilities and utilities require high animal throughput for solvency. High animal throughput requires high-energy rations that are more directly edible by humans. . . . With fossil energy no longer cheap or plentifully available, and no energy panacea to replace fossil energy, can we subsidize meat production by large energy infusions to keep grain-fed meat a part of our American Heritage?
—W. L. Roller, H. M. Keener, and R. D. Kline, *Energy Costs of Intensive Livestock Production* (St. Joseph, Mich: American Society of Agricultural Engineers, June 1975), p. 8

THE LIFE OF ALL LIVING THINGS—plant or animal—depends on a basic chemical process called photosynthesis. This process occurs in the cells of plants where, with the aid of energy from the sun, complex organic molecules needed for growth and other plant processes are built from carbon dioxide, water, and other simple molecules. Each time molecules are bonded together by photosynthesis, the bond becomes stored energy. When an animal eats plant material, digestion and other bodily processes break down these large molecules into nutrients and energy for its growth, body heat, movement, and other animal functions. Both plants and animals are able to build up a surplus of large, energy-laden molecules. Animals store theirs in the form of fats; plants store theirs in the form of proteins, starches, and oils that are usually most concentrated in seeds, nuts, and grains.

Agriculture is essentially human effort aiding the conversion of solar energy into food. Grain is agriculture's standard form of energy. Industry has its barrels of oil, agriculture its bushels of corn. We should prefer a mode of agriculture that is neither wasteful of this energy nor destructive of the environment.

71

Energy-Rich Factory Feeds

At present most plant materials produced on American farms are run through animals to produce meat, milk, and eggs. We feed about 90 percent of our corn, oats, barley, and sorghum and over 90 percent of our unexported soybean crop to animals.[1] Steadily rising demand for animal products over the years has forced more and more land to be devoted to raising corn, soybeans, and other grains. Grain is easily stored, ground, mixed, and measured and is more easily processed than the rest of the plant. Whether in whole kernels or ground and mixed with additives, grain can be conveniently pumped through the factory's automated feeding machinery.

Factory poultry and pigs must be fed rich, easily digestible grain concentrates because they have digestive tracts that cannot break down rough grasses, hay, and silage. The broiler industry boasts of the efficiency with which its fast-growing strain of bird converts grain to meat; but this dubious efficiency is achieved by using the highest-calorie grains. As we shall see, it is not really an efficient use of these grains at all. Less efficient still is the use of grain to fatten cattle, which can digest grass and other foods not digestible by humans. Because beef-factory farmers are paid by the pound, their production goal is to produce the heaviest animal in the shortest possible time. Feedlots are, in that sense, efficient factories. But because of the grain-rich diets and no exercise, they are more fat factories than protein factories. A "choice"-grade beef carcass from a "well-finished" feedlot steer has about 63 percent more fat (and more calories and cholesterol) but less protein than a "standard"-grade carcass from a grass-fed animal.[2] Then, at the packing plant, supermarket, and kitchen, grain and energy are wasted when this fat is trimmed away.

Factory Efficiency?

Animals have biological limits on their ability to convert plants into other material. A pound of grain cannot produce a pound of edible protein because not all of the grain is digested into nutrients. Not all of these nutrients are absorbed into the bloodstream, and not all of those absorbed into the bloodstream are converted into muscle. Some of the nutrients are burned up for movement, cell replacement, internal processes, and other nongrowth functions.

Not all of the animals that we feed end up on our tables. Unlike machines of metal or plastic, animals are perishable, and this perishability is a substantial factor in the cost of putting meat, eggs, and milk on American tables. After investments of time, energy, and resources, an animal—so industriously prepared into a walking food package—can be quite uncooperative: it can get sick and die. A study of Missouri pig farms

A Texas feedlot.

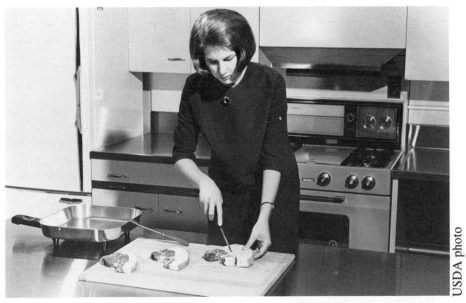

Denied exercise, factory animals fed high-calorie meals produce fatty flesh like these pork chops.

revealed that over one third of all pigs born on the farm die there as well.[3] Nearly two of every ten calves die in veal and dairy factories, and losses can run as high as half the calf "crop."[4] Life is so hard in the cages of the automated layer house that about 30 percent of hens put in the cages die or are "culled" during an average eighteen-month laying cycle.[5]

Animal behavior, as factory farmers are only too aware, also adds to the unreliability of animals as protein machines. They don't always mate, breed, and turn out offspring in clockwork fashion. A missed heat period, a failure to conceive, or a miscarriage means that months of feed invested in the breeding animals brings no salable pig or calf, and no return. One pig expert estimates that each pig lost at birth represents a waste of about 130 pounds of feed used in feeding the breeding herd.[6]

If we are to analyze the real efficiency of animals as food machines, then, we must add in all grain and other food energy spent in rearing and maintaining breeding animals and all losses resulting from infertility and deaths. When these are figured in, only about 17 percent of the usable grain or food energy fed to a dairy herd is recovered in milk, and only about 6 percent of that fed to a beef herd is recovered in edible meat.[7]

But animals take in more than mere calories of energy from grain and feedstuffs. They also consume protein and other nutrients that could be used to nourish humans. Protein is poured through animals and lost at an astonishing rate. The accompanying table shows that animals are not very efficient at producing protein for us from the protein we feed to them. The figures are based on assumptions about fertility, mortality, reproductive rates, and production costs usual in farming systems. The figures for protein refer to the whole animal; the percentage for the edible portion would be even smaller.

Percent of Feed Protein Converted to Animal Protein

Animal	Percent
Dairy cow (milk)	22
Hen (eggs)	23
Broiler	17
Pig	12
Beef animal	4

Source: J. T. Reid, "Comparative Efficiency of Animals in the Conversion of Feedstuffs to Human Foods," in *New Protein Foods,* eds. Aaron M. Altschul and Harold L. Wilcke, vol. 3 (New York: Academic Press, 1978), pp. 116–143.

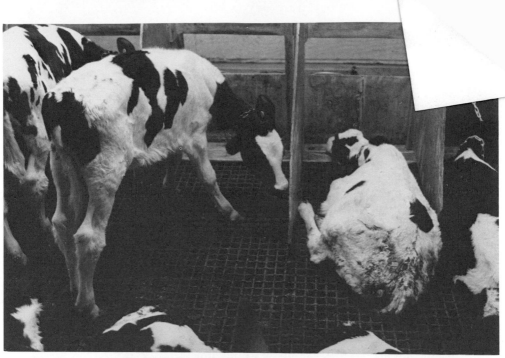

Chances are that one of these four newborn dairy calves will die soon after being brought to this confinement nursery.

"Energy and protein are lost when feedstuffs are converted to meat and milk."—
D. G. Fox and J. R. Black, "New Tool Pulls It All Together—Systems Analysis," *Confinement*, May 1976, p. 12.

The High Cost of Waste

The waste of grain and resources because of animals' inefficiency as converters is costly, but it becomes even more so in light of the unpredictable nature of animal production. Disruption of production schedules causes dollar losses to producers and consequently higher prices paid by consumers. Stress and disease, even when they do not cause death, make factory animals more expensive to prepare for market. In poultry, for example, lowered productivity from one disease alone (coccidiosis) adds $88 million to the feed bills of broiler producers each year.[8] According to the Livestock Conservation Institute, an industry group that specializes in losses associated with animal production, pig producers lose more than $187 million each year from dysentery, cholera, abscesses, trichinosis, and other swine diseases.[9] In cattle, brucellosis costs producers $30 million in lost milk production and dead calves each year.[10] In all, it is estimated that U.S. livestock and poultry producers lose some $3 billion each year because of diseases and parasites.[11] This amounts to over 7 percent of *gross* farm income from all U.S. animal and poultry production and adds substantially to the animal products food bill.[12]

Grain and other investments in animal production are wasted when animals are injured or killed in shipment from farm to farm and to the slaughterhouse. Increasing specialization in the cattle and pig industries has separated operations; there are breeders, producers of young calves and pigs, and producers who "finish" the animals to market weight. Cattle and pigs may be moved several times from farm to auction to farm and back again. Most animals are moved in crowded trucks and subjected to stresses from exposure in cold weather or overheating in warm weather. Crowding, jostling, and rough handling cause bruises and other injuries. Not counting poultry, losses related to transportation amount to about $15 million each year from dead and crippled animals and another $46 million from carcass-damaging injuries and stress-related deaths. For cattle alone, $102 million is spent on last-resort medication.[13] Factory animals are especially vulnerable to shipping losses; according to an expert, factory pigs are "more delicate than open-raised pigs. . . . They're not only tailless and cleaner, but much more nervous" and prone to stress-related deaths and injuries.[14]

Then there are the wastes at the slaughterhouse—losses from the human food chain because many animals and parts of animals are damaged or diseased. Under federal meat and poultry inspection (which accounts for over 90 percent of all animals inspected), animals are examined before slaughter for obvious signs of disease or abnormality. Each year about 116,000 mammals and nearly 15 million birds are condemned before they enter federally inspected slaughterhouse doors.[15] Inside, after killing, inspectors weed out another 325,000 whole carcasses

Veal calves' carcasses.

USDA photo

and over 5.5 million major parts (heads, legs, shoulders, etc.) of cattle, calves, hogs, and sheep each year.[16] At federally inspected processing plants, inspectors condemn about 140,000 tons of poultry meat annually.[17]

Factory methods add to this mountain of discarded flesh. The feedlot liver syndrome discussed in chapter 2 results in a great number of carcasses condemned at the slaughterhouse; liver condemnations alone amount to a $10 million annual loss to producers.[18] Nervous, "meat-type" factory pigs that drop dead at the slaughterhouse end up in the pile. Marek's disease—cancer—in chickens is the leading cause of condemned carcasses at processing plants; it is believed to be related to the stress of crowding and the supergrowth of birds in modern broiler factories.[19]

These tonnages of condemned flesh never become the food they were intended, nor does much of each animal that does pass inspection; heads, feet, and other inedibles amounting to about half the animals' live weights go with the condemned flesh into fertilizer, soap, and other by-products.

Some of the flesh pile goes into protein supplements that go back to farms to "beef up" another round of animals. The animal factory as food factory is wasteful indeed. The meat industry tells us that nothing is wasted at the slaughterhouse, but all this really means is that they make extra money selling all of the nonedibles. It ignores the waste inherent in producing the whole animal in the first place.

Fuel for Factories

Agriculture, primitive or modern, has "ancillary" energy costs, that is, the energy or power used in manipulating the flow of solar energy through plants to the final food product. Energy must be spent in clearing, cultivating, sowing, weeding, harvesting, separating, and storing food—grain, let's say. Energy is spent in moving, grinding, and mixing this grain and in feeding it to, watering, and otherwise caring for the animals who eat it. Energy is spent in shipping, slaughtering, processing, packing, storing, and cooking the animal food products. We even spend energy in advertising these products.

Over the years, agriculture—American style—has moved away from human labor for these tasks and toward machine power. In the late nineteenth century, when resources for power and machines were cheap and wages were rising, agricultural planners began to choose machines for their convenience, control, and low expense relative to human labor. The history of the poultry industry sketched in chapter 1 illustrates this choice toward the capital side of the trade-off between capital and labor in production. As a result of thousands of such choices over the years, we have evolved an agricultural technology that has increased overall

Energy use: forced ventilation for an egg factory.

Energy use in a factory broiler house: the large disks are gas heaters. Automatic waterers hang from hoses and the two pipes running the length of the building are automatic feeders.

productivity per farmer, but only by huge investments of energy from petroleum, coal, and hydroelectricity. American agriculture uses more energy in production than it puts out in the form of food. An American scientist, David Pimental, estimates that if the whole world were fed diets of food produced by U.S. agricultural technology, known petroleum reserves would be exhausted in thirteen years.[20]

Animal factories are especially gluttonous of energy. Energy is spent in manufacturing the buildings, hardware, and supplies for confinement systems; it is used to keep the machinery going and to maintain the "controlled environment."

In addition, the placement of animals in confinement creates problems that require energy for solutions. For example, pigs and poultry have more difficulty than some warm-blooded animals in maintaining body temperature. They easily get too hot or too cold. On a traditional farm, they could keep warm in cold weather by nestling in bedding placed in shelters. In hot weather, they could cool off in shady, damp soil. In the factory, however, when the environment becomes uncomfortable, the operator must use energy to adjust it, otherwise productivity falls. Large factory buildings packed with animals must have very powerful ventilators to keep foul air moving out and fresh air in. Sprayers, foggers, and, on some farms, air conditioners must be used in hot weather to keep crowded

animals from overheating. Energy must be spent in moving feed in and wastes out. And waste must be treated if the farmer wants to avoid pollution problems and neighbors' complaints. At one Kansas pig factory, for example, where 7,500 pigs produce 30 tons of fresh manure each day, the farmer runs up $90-a-day electric bills running odor-control agitators in his waste pits.[21]

Heaters are necessary if young animals are to survive separated from the warmth of their mothers' bodies and dry bedding. Young animals are vulnerable to chills, especially on drafty, cold concrete or metal-slatted floors. To improve survival rates, farmers use electric heating pads and infrared lamps in pens and stalls. Even older, growing animals get chilly in factory buildings and when they do, they eat more and burn up more energy to maintain body heat. Farmers who want rapid weight gains on schedule have two choices, both costing energy: either turn up the heat or feed more grain.

More Waste

When the energy spent in running animal factories is compared with the protein and food energy they produce, the ratios show waste again. The food energy we get from pig factories is only about 34 percent, and from broiler factories only about 14 percent, of the fossil fuel energy used in

Energy use: artificial daylight for faster broiler production.

production.[22] By way of contrast, the soybean and corn crops fed to these animals produce an energy dividend: corn produces nearly seven and soybeans nearly six units of food energy for each unit of fuel energy used in production.[23]

Now let's look at protein produced for each unit of fossil fuel used in production. If the corn and soybeans consumed in the pig and broiler factories were consumed instead by humans, for each unit of fossil fuel energy used in production, we would get back nearly five times the protein produced by either the pigs or the broilers.[24] According to animal scientists at Ohio State University who have been studying energy costs in factory systems:

> [These data] forcefully bring to mind one obvious fact—man certainly doesn't grow animals to amplify our food energy availability! No knowledgeable person ever thought we did. Even the best of the animal enterprises examined returns only 34.5% of the investment of fossil energy to us in food energy whereas the poorest of the 5 crop enterprises examined returns 328%.[25]

Agriculture in general may be the process whereby solar energy is converted into food, but animal factories are something else. They are, indeed, protein and energy factories in reverse. They do more to waste protein and energy than to increase it.

6 Hidden Running

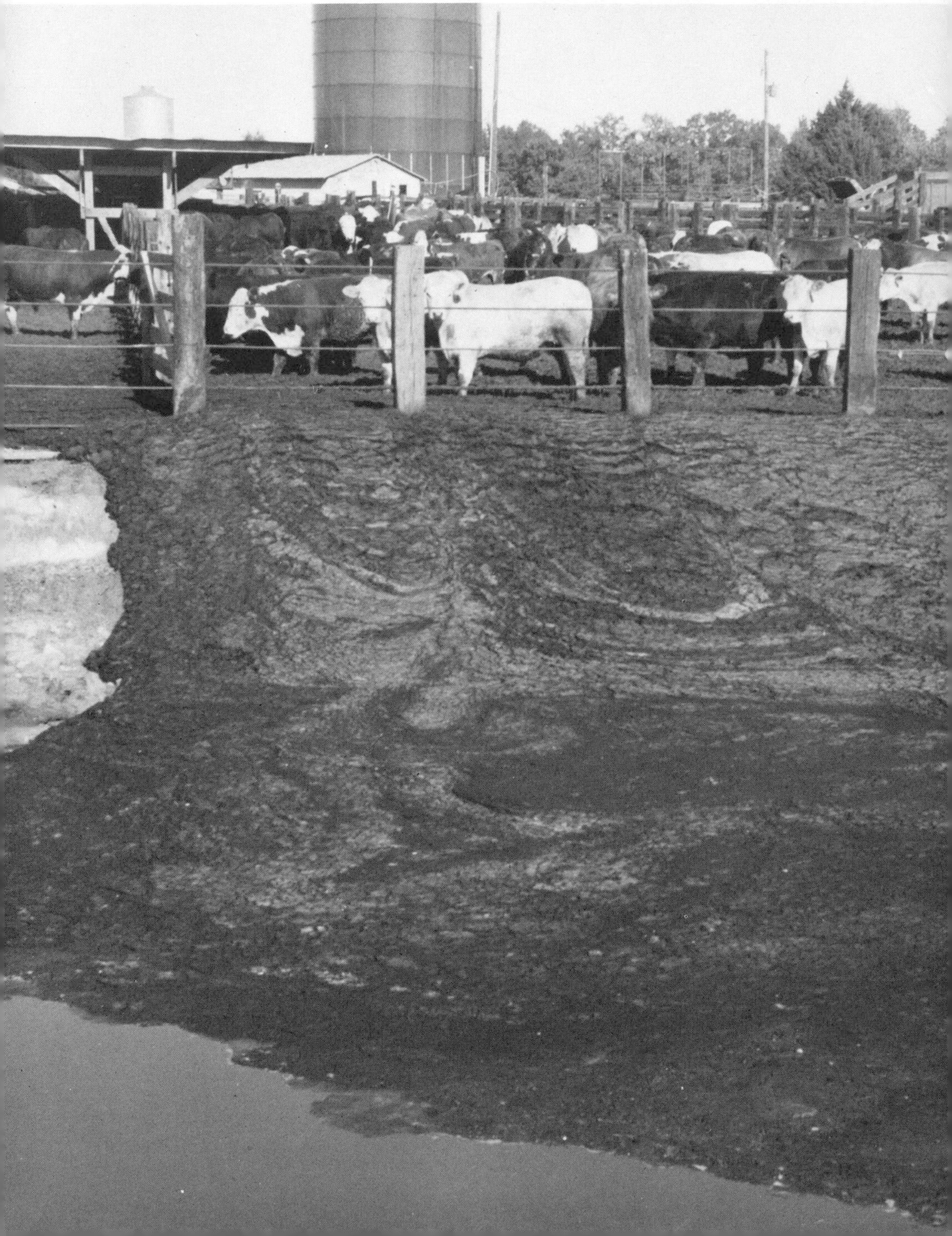

COSTS OF THE FACTORY
Who's Paying Them?

In recent times, the ill-defined "balance" of nature has dipped dangerously and the environment has deteriorated alarmingly. One aspect of this deterioration is pollution from intensive animal production.
—H. A. Jasiorowski, "Intensive Systems of Animal Production," *Proceedings of the III World Conference on Animal Production,* ed. R. L. Reid (Sydney: Sydney University Press, 1975), pp. 369, 383

HUMAN ATTEMPTS TO MAKE animal agriculture efficient through factory farming cause unexpected problems. Some of these problems are not easily overcome, and impose continuous risks and costs. We have already noted the increased risk of far-reaching accidents in the food chain like the PBB disaster in Michigan.

There is another dangerous new environmental problem caused largely by factory farming's extensive use of antibiotics: diseases are returning in virulent new forms because bacteria are developing resistance to these drugs. When someone gets sick from these bacteria, the familiar antibiotic "wonder drugs" are useless against them. Diarrhea, septicemia, psittacosis, salmonella, gonorrhea, pneumonia, typhoid, and childhood meningitis are a few of the diseases that have developed strains resistant to penicillin and other antibiotics.[1] Since about half of all U.S. antibiotic production is used as additives in animal feeds, experts believe this practice is a major contributor to the rise of the new antibiotic-resistant diseases.[2] Coliform and salmonella, two of the types of bacteria known to have resistant strains, are major contaminants of meat and poultry in the supermarket. The distribution, sale, and consumption of these products provides a wide channel for the flow of antibiotic-resistant bacteria into humans and the environment. Of course, animal factory experts and drug companies take the position that all of this is "just a theory"; and they continue to fight proposals for regulations.

83

I've seen the U.S. situation, and would not swap our setup for yours. You use such massive amounts of antibiotics that if you dropped them, you would have a catastrophe.

—R. Wilmore, "Many Problems in British Antibiotic Policy," *National Hog Farmer,* October 1975, p. 28

A less obvious problem caused by factory farm systems is the costly disruption of the complementary relationship between plant and animal agriculture. On traditional farms, most animals are unconfined and can disperse their wastes over the land with no detrimental effects. Odor problems are nonexistent because the wastes either dry or dissolve into the soil. Cycles for the exchange of energy and nutrients among soil, plant, and animal are continuous without risk of pollution and without additional inputs of labor and energy. On factory farms, animals and land are separated and these cycles are interrupted, with the result that restoring nutrients to the soil has become expensive and time-consuming. On the factory farm, animal wastes pile up rapidly and may pollute air, ground, and surface water and attract hordes of rodents, birds, insects, and other "pests."

Factory Sewage

Animal agriculture is the greatest producer of sewage wastes in the United States. A 60,000-bird caged-layer house produces about 82 tons of manure every week. In the same period, the 2,000 sows, boars, and feeder pigs in a typical factory produce nearly 27 tons of manure and 32 tons of urine.[3] In all, our farm animals produce about 2 billion tons of manure each year—about ten times that of the human population—and half of this comes from confinement operations.[4]

Not all animal wastes are the same. Animals fattening in factories are fed such liberal amounts of rich feed that they absorb only a fraction of the nutrients. Their wastes contain more protein, organic matter, nitrogen, phosphorous, and other material known to cause pollution problems than do the wastes of animals on normal diets.[5] Feed additives used in factories—antibiotics, copper, and arsenic, among others—make the manure a more potent pollutant. So the trend toward factories has brought us not only the new problem of dispersing waste, but a new kind of waste that is more polluting than ever.

If animal wastes reach ground or surface waters, they take up oxygen as they decompose. Water overloaded with wastes becomes stagnant and incapable of supporting fish and other animal life. Waste from duck factories on Long Island, New York, has killed shellfish and caused odors

Droppings under egg cages.

that interfere with public recreation.[6] During the 1960s, runoff from Kansas feedlots made that state responsible for half of the reported fish kills in the entire nation. Pollution from animal waste was responsible for 82 percent and 99.5 percent of the fish killed in Kansas in 1964 and 1965, respectively.[7]

The Factory as Neighbor

Factory farms stink. Up close, odors from wastes can be literally overwhelming. On first visiting factory farms—before our noses became "calloused"—we felt fatigued and irritable for hours afterward.

Attempts to control odors with chemicals are being explored. There are chemicals to halt the biological breakdown process that releases odorous gasses, chemicals to mask odors, chemicals to counteract one or more of the odorous components in waste, and just plain old deodorants.

A pig farm with waste treatment ponds.

Chlorine, lime, potassium permanganate, and hydrogen peroxide have been used to hold down odor problems. In all, some twenty companies put out as many products supposedly helpful in controlling odor problems. In recent tests at the University of Illinois, none of these significantly reduced odors.[8]

Factory wastes attract sparrows, starlings, and insects, especially flies. In a letter to a farming magazine, one pig farmer complained: "I have a total confinement system that is eaten up with flies and I know it's a health hazard."[9] Complaints about excessive flies and rodents have been a major cause of litigation between factory farmers and their neighbors. These "pest" animals can transmit diseases to and from farms. But what problem can't be solved by one more feed additive? Sure enough, commercial additives have become available that contain hormones and toxins to hold down bird and insect populations that would otherwise feed on the matter in animal wastes.

Not surprisingly, there has been an increase in the number of lawsuits against factory farm owners. Neighbors of a Missouri cattle feedlot and pig factory claimed that runoff damaged ten acres of land, polluted two ponds, lowered milk production in their dairy herd, and caused the deaths of six cattle. A Kansas pig factory was forced to shut down because of complaints from neighbors to state authorities. In Minnesota, protesting neighbors and townspeople blocked construction of a large pig factory under local ordinances when the builder could not assure them that there would be no air or water pollution.[10]

As a result of the increase in pollution problems and the skirmishes between factory farmers and their neighbors, state and federal authorities have begun to step in. Several states require producers to control their wastes or treat them to contain odor and pollution problems. The federal government's Environmental Protection Agency put forth a proposal in 1976 that would have regulated some 3,200 of the nation's 700,000 beef, dairy, and pig "feedlots." A court decision later the same year expanded the scope of coverage to about 95,000 operations. These rules, however, prevent only discharge of untreated wastes into navigable streams; they have not effectively ended the problems of odors and pests.

Solutions to waste problems lag behind the development of factory systems and add to the complications of farming. If farmers want the maximum fertilizer value from animal wastes, they must devote much time, labor, and energy to hauling and spreading the stuff on fields. But frequent field application is bound to cause odor problems and complaints from the neighbors. There are times, too, when field application is

The concentration of animals like these pregnant sows takes much of the labor out of pig production, but the huge volume of waste produced creates serious pollution problems.

impossible, such as during the crop's growing season and when the ground is frozen or wet. Moreover, heavy tractors and equipment compact the soil, making it less able to absorb the manure and increasing the potential for pollution. An additional problem for some factory farmers is that they do not have enough land to safely assimilate all of the wastes their animals produce.

Farmers are in a bind. High land costs, lack of land availability, and other factors force them to concentrate animals in factories; but then they need additional facilities to get rid of the daily pile-up without causing pollution problems. As one pig producer states the dilemma: "We've got two routes to go. Either disperse the livestock so we don't have to disperse the waste or concentrate the livestock and have problems dispersing the waste." [11]

Whether farmers decide to store, disperse, or degrade animal wastes, the facilities can be expensive and a chore to operate. Some factory storage facilities use forced air, augers, or pistons to push waste from factory buildings to tanks or holding ponds. Some farmers gouge out large, pondlike "lagoons" to hold wastes where bacteria digest everything into simple, nonpolluting elements. These anerobic lagoons are the most labor free, but they can give off strong odors. If odors are a problem, farmers must pump in fresh water or use pumps and aerators to stir in oxygen.

The loss of nitrogen from animals' wastes is perhaps the most costly aspect of factory disposal systems. Field crops need nitrogen, and to provide it American farmers spread about 10 million tons of commercial fertilizer each year at a cost of about $2.5 billion. Livestock and poultry produce enough waste each year to provide $1.5 billion worth of nitrogen fertilizer, but because factory farm methods of waste collection and disposal destroy nitrogen, only about half this amount is available for fertilizer. On the largest factory farms, intensive waste treatment methods cause even greater losses of nitrogen. In lagoons, for example, as much as 80 percent of nitrogen is destroyed by microbial and chemical action.

Most factory farmers aren't concerned about the efficiency and the greater ecological good of nutrient recycling. It is just too much trouble and too expensive to get nutrients back to the soil. One such farmer's comment is typical: "Until fertilizer gets more expensive than labor, the waste has very little value to me." [12] Such is the economic logic of capital-intensive American agriculture.

Public Subsidy: Disease Control

Animal farming is more costly than growing plants because mammals and birds are susceptible to more diseases than plants are. And because they are closer to humans in evolution, these animals can harbor diseases transmissible to humans. Trichinosis, brucellosis, leptospirosis, and salmonellosis are just a few of the eighty-odd bacterial, viral, and parasitic diseases that we can catch from farm animals.[13]

Because of these threats to human health and the perceived (perhaps once real) "need" for animals in the human food chain, animal disease control programs at public expense began in the early 1800s. By about 1880, Congress established the Bureau of Animal Industry within the U.S. Department of Agriculture "to provide means for the suppression and extirpation . . . of contagious diseases among domestic animals."[14] The bureau established a disease control scheme that used quarantine, killing, and burial or burning of entire herds and disinfection of premises in attempts to "eradicate" all of the disease-producing organisms within the United States.

Eradication campaigns are costly; they require veterinarians and other expert personnel and they provide for indemnity payments to farmers whose animals must be destroyed. Since 1934, when the campaign against brucellosis began, over $1 billion has been spent trying to wipe out this disease in U.S. cattle.[15] Efforts to halt just this one disease are costing the federal government about $33 million and the states another $25 million each year, and the end is not yet in sight. Lately, this eradication program has bogged down and some experts believe that if eradication is possible at all, the cost may be prohibitive.[16] Hog cholera was believed to have been nearly eradicated until the spring of 1976 when an outbreak in eastern states sent government exterminators out with their emergency gear. They killed and buried 24,000 hogs, and farmers were paid $3 million in indemnities. Then, in January 1978, the disease was pronounced eradicated by the Secretary of Agriculture.* Total expenses to "eradicate" hog cholera came to over $140 million of public funds.[17] Similarly, when exotic Newcastle disease struck poultry operations in southern California in the early 1970s, government disease control squads went to work. Twelve million destroyed birds, $56 million, and a year and a half later, the disease was adjudged "eradicated," although small outbreaks were reported in each of the next three years.[18]

As new animal diseases emerge, new eradication programs follow. Pseudo-rabies in pigs, for example, has been taking virtually entire herds on some farms in the Midwest since 1973. Representatives from USDA, the National Pork Producers' Council, and other groups have proposed a

---

* *Eradicate,* as disease control authorities use it, is a technical term and their definition of it does not mean that the disease-causing organism has been made extinct worldwide. It just means that its movement has been brought under control so that no new cases of the disease are reported in the United States.

five-year eradication program that would cost $44 million.[19] Others estimate that the program could cost state and federal governments as much as $90 million.[20] A new disease, African swine fever, has reached the Western Hemisphere and, because there is no vaccine and no treatment, government and industry are already discussing an eradication program. If the disease takes hold in the United States, experts estimate that the cost of a five-year program could reach $290 million.[21]

Of course, we aren't arguing that it would be better to let the animals die and save the money. We are simply trying to point out the irrationality in our practice of maintaining huge populations of food animals. These reservoirs for diseases will continue to generate higher food costs— whether from rising costs of control programs, veterinary services, and medical supplies or from wasted grain and other investments when the animals die. The choice is not between animal deaths or disease control; we could just as easily reduce our reliance on animal agriculture with its costly problems.

Although disease control programs were conceived of and have been justified on the grounds of public health, many of the diseases under control measures have no effect on humans. Rather, the control programs pay out public funds to reduce the true cost of animal production. They constitute a public subsidy to farmers and the meat industry.

Public Subsidy: Milk Price Supports

Like the meat and poultry inspection program, the milk price support system is another publicly funded prop for the animal industry. While the inspection program keeps meat prices artificially low by covering the cost of inspection out of taxes, the milk price support system keeps consumer milk prices artificially high. Under this program, the federal government sets minimum prices that must be paid by processors to milk producers. To guarantee that all milk sells at these prices, the U.S. Department of Agriculture buys all unsold (surplus) milk at announced prices; this system keeps farmers in the dairy business when they would otherwise be producing other crops. The "benefit" to consumers is that they have more than enough milk to go around. So much, in fact, that 135 million pounds of butter, 42 million pounds of cheese, and 338 million pounds of nonfat dry milk (products equivalent to 3.2 billion pounds of fluid milk) had to be removed from the market by the government during the 1977/78 marketing year and added to inventories from the previous year.[22] After donations to foreign and domestic food aid programs, sales to the army, and sales of nonfat dry milk for use as a protein supplement in animal feeds, the government had a total of 819 million pounds of butter, cheese, and nonfat dry milk in storage at the end of the 1977/78 marketing year.[23]

The program persists because of the raw political power that dairy

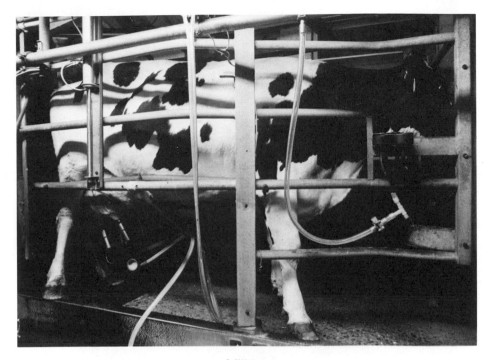

Milking.

industry money can buy. In 1976, the dairy lobby contributed over $1.3 million to the campaigns of House and Senate candidates—second only to the most notorious big spender, the American Medical Association.[24] And could the $126,000 campaign contribution from the dairy lobby have influenced Jimmy Carter's decision to raise the level of federal milk price supports by 11 percent, adding $1.2 billion to retail milk prices?[25] In all, estimates of the annual cost to the public of this subsidy program range from $400 million to $1 billion.[26]

Public Subsidy: Factory Hardware

Government tax policy subsidizes the factory approach to animal farming and promotes specialization at the expense of diversity and versatility in farm systems. It promotes hardware rather than human inputs to production. In many states, for example, portable confinement units are not considered buildings or real estate and are not subject to real property taxes. More importantly, highly specialized pig and poultry factories and hardware now qualify for the "investment credit" deduction from income taxes. This costs the public $10 million a year in lost taxes. All-purpose, conventional farm buildings do not qualify. Farmers who

Some of the specialized hardware in a modern egg factory that entitles its owners to an "investment credit" deduction from their income taxes.

maintain their old-fashioned barns and animal shelters must do so without the tax advantages their factory-building neighbors now have.[27]

Federally guaranteed loan programs have been misused by officials to fund large cattle and pig factories. Testimony at recent hearings before Senate committees revealed that the financing practices of the Small Business Administration and USDA's Farmers Home Administration favored factory systems to the economic disadvantage of family farms, in violation of the Food and Agriculture Act of 1977. SBA set its own size standards that permitted funding of operations with up to $1 million a year in gross sales, and was routinely financing pig factories in the top 1 percent of sales volume.[28]

Public Subsidy: Factory Designers

Who pays for the experts who are leading agricultural technology toward the use of more hardware and energy, employment of fewer and fewer people, greater specialization and vulnerability, and monopoly control of food production?

We do. The bulk of agricultural research in the United States is conducted by the tax-supported "land grant" college complex and the U.S. Department of Agriculture. Composed of 69 state universities, their colleges of agriculture, experiment stations, and extension services, this complex employs over 35,000 people at an annual cost of $600 million.

This system was set up to serve the needs of consumers, family farmers, and farm workers; but one of its main jobs now, it seems, is providing expertise to the largest agribusiness and food industry corporations. Public funds supported a Cornell University project to develop mechanically deboned poultry "meat"—pulverized bone, cartilage, and flesh—so that industry could put it in hot dogs and luncheon meats and fetch a better price for slaughterhouse leftovers. Scientists at the University of Georgia used public funds to perfect a mechanical broiler harvester that herds birds onto a conveyor belt and out to the trucks. At the request of a Georgia broiler company, the university's poultry science department started a program of offering workshops to a single company's management personnel who, in turn, "tailor the program to company grow-out programs." [29] In other words, the university provides a publicly supported consulting firm for Georgia's broiler companies.

Many government and university scientists act as publicly supported inventors for industry. They develop a drug, machine, or system and then allow a company to turn it into a commercial product. Much of factory farming technology is developed in this way, and publicly supported scientists have no qualms about it. [30] But they assure us that their tax-supported work to help agribusiness companies brings progress and improvement for everyone. Under such a system of technological development though, any benefit to farmers and consumers is coincidental rather than intentional. Agribusiness, not farmers or consumers, makes the vital decisions about what subjects should be investigated and what direction technology should take.

In the process, "scientific objectivity" has vanished and the public has lost the benefit of expertise in finding the best technology for all society. The trend toward animal factories is highly touted for production efficiency which, it is argued, keeps prices of animal products from skyrocketing. But there are hidden costs in this illusory efficiency. If the true cost of factory farming were known, people might not be willing to subsidize so expensive a mode of food production.

AGRIBUSINESS
The Farmer as Victim, or Who's Making the Real Money?

*I think we need a systems-type research program in agribusiness
. . . the kind that gave us the atomic bomb and put a man on the
moon.*
—Dr. John H. Davis, coiner of the term
agribusiness, quoted in "Agribusiness
Needs of the Future," *Broiler Industry,*
September 1976, p. 26

IN RESPONSE TO CONTINUING criticism of policies favoring agribusiness, the U.S. Department of Agriculture and agricultural establishment leaders reply that most of the nation's 2.7 million farms are still "small" and family run. These farms make up about 95 percent of all farms and account for 65 percent of all farm products sold when all types of farming are lumped together.[1] But these simplistic statistics ignore the deeper realities of American agriculture's domination by "agribusiness." If we look at specific crops and how they are produced and marketed, we can see that big agribusiness farms often overwhelm small, independent farmers. In 1975, 95 percent of vegetables, 85 percent of citrus fruits, 98 percent of milk for drinking, 97 percent of broilers, 54 percent of turkeys, 40 percent of eggs, and 22 percent of fed cattle were produced either by agribusiness corporations and cooperatives or under contract to them.[2] About 89 percent of all food and fiber is produced by 35 percent of our farms and ranches, and each of these farms maintains an average of about $.5 million worth of land, buildings, and machinery.[3] Our very largest farms make up only about 6 percent of all farms, but they get 53 percent of all cash receipts from farming.[4]

America's remaining independent farmers are caught in a current of economic developments that will either throw them off the farm or propel

One manufacturer's offering of hardware for the factory production of pigs. Used by permission of the Clay Equipment Corp., Cedar Falls, Iowa.

them into bigness and specialization. A Kansas farmer who farms land that has been in his family for a century explains how these developments force farmers into a dilemma.

> The returns to capital and labor invested in production agriculture are so low (and steadily growing smaller) that increased volume and specialization is the only alternative to a declining standard of living for farm people. The farmer who decides to stay small and keep farming has just initiated creeping bankruptcy.[5]

Bigger and Fewer

Prices of farmland are rising because of demand for land from nonfarm interests and from established farmers with mortgageable land and equipment who want to expand. Big farms are getting bigger, the total number of farms is declining, and all farmers must make an effort just to stay in farming. The big farmers increase farm productivity to cover the cost of added acreage; the small farmers, cut off from additional acreage that could improve their livelihood, must wrest even more from their land to make ends meet—if they want to remain farmers. At the same time, production goals get more elusive because the labor pool available to farmers is drying up; fewer and fewer people are skilled and reliable enough to do the increasingly complex kinds of work needed on modern farms—at wages farmers can afford.

Changes in marketing of farm produce reinforce the trend to large-scale farming methods. Traditional local markets for farm products have been displaced by bigger, regional markets. In the interests of efficiency and, ostensibly, of health and sanitation, many states now require inspection of produce. Officialdom and agribusiness prefer to conduct inspections, standardized measurement procedures, and processing at a few large centers rather than at hundreds of small ones all over a state. In many states, health regulations and milk-processing companies require farmers to have special barns, equipment, tanks, and coolers that small farmers cannot afford.

The Bias toward Agribusiness

Farmers are surrounded by a bias in favor of big, specialized, capital-intensive farming. For years, government policies have favored size.[6] Only 16 percent of farms—the very largest—received almost half of the $530 million in tax-supported assistance paid to farmers in 1975.[7] In its 1975 census, USDA decided that a farm is not a farm unless it is an "economic unit" that sells more than $1,000 worth of agricultural products a year; by definition, 570,000 farms ceased to be farms.[8] The biggest operators see no need for small farms either; some, in fact, see small operations as a threat. The commercial poultry industry is becoming concerned about "peripheral birds"—backyard chickens, small flocks, and other birds—because these birds can carry diseases that can wipe out their more vulnerable large commercial flocks.[9] One California poultry expert warns industry people of the "ever-increasing political strength" of the keepers of "peripheral birds."[10]

Most pervasive of all is the bias toward agribusiness by farming magazines, agricultural college extension agents, and salespeople from the companies that supply products to agriculture. Together, these elements put forth an ideal of farming that represents their view of profitable or successful farming. The progressive farmer, the "pork all-American" or milk production champion, is the one who goes all out for production and uses anything and everything to get it. Implicit in this ideal is the notion that the farmer should go into debt to acquire the latest in factory buildings or equipment. To help farmers make expensive land pay, one farm magazine advises: "Consider intensification. . . . At the present time, you can put in improvements and expand livestock cheaper than you can buy more land."[11] The magazines, of course, are loaded with advertisements from manufacturers of buildings, equipment, drugs, and supplies used in factory systems.

The advertiser's message. It gets through to farm magazines' publishers, too.

The magazines tend to be hostile to alternatives to the agribusiness way of food production. When USDA recently proposed a total of $3 million (only about .02 percent of its budget) for two projects to help stimulate local production (one would have helped promote roadside markets, the other would have encouraged community gardens in urban areas), both programs were blasted in *National Hog Farmer*. The editor wrote: "Why don't we just turn the Department of Agriculture over to the do-gooders?" [12] When a Washington, D.C.–based group, The Exploratory Project for Economic Alternatives, set forth a plan that would aid small farmers with subsidies and other incentives, *Beef* magazine said that "the best thing that can happen to this report is that it be consigned to molder away in some file cabinet and forever be forgotten." [13]

Animals: The Perfect Cash Crop

Farmers now have a steady flow of expenses; they have payments for land, buildings, machinery, fuel, and, of course, living expenses. They must produce some crops that generate a steady flow of cash. Cash income from field crops is one solution, but these crops are dependent on market and weather uncertainties and are slow and seasonal in coming. Animal production has always been the main source of cash flow on many farms because it provides a steady flow of saleable commodities.

Today's day-to-day cash requirements can hardly be met by traditional methods of animal farming, however. Thus begins the capital-expansion spiral into factory systems and methods. First, flock or herd size must be increased substantially if more money is to be made. But then labor requirements impose limitations unless the farmer can afford to invest in factory buildings and equipment. The expense will be prohibitive unless the farmer is willing to maintain a very large number of animals and push them through with factorylike speed and efficiency. As one expert puts it, "With a $1,000 investment per sow you want to get as many pigs out of the buildings as possible. It's a far cry from a $50 individual house and some fencing used in the past."[14] The whopping costs of capital intensification of animal production, then, are spread over the herd or flock. To ensure profitability, the farmer must employ the full arsenal of factory methods, including crowding, speeding cycles, use of drugs and growth promotants, and the other factory management tools. Farm animals have been put to work as never before in the modern factory farm.

Factory Farm: Labor Efficient?

Factory systems are not at all cheap. Estimates for setting up pig-confinement facilities range from $1,000 to $1,500 per breeding sow.[15] The cost of setting up a family-size dairy farm can easily run about $200,000. These estimates are for the cost of confinement buildings and equipment—none includes the cost of farmland.

To keep their heads above water, then, farmers must devote more time and energy to farming than ever before. This is the irony of "labor-efficient" factory farms: the financial burdens are so great that farmers must work harder than ever to meet payments. Farmers now go to night school to learn stockbroker tactics like "hedging" and to understand futures trading and other market intricacies. Their headaches are worse than ever and many farmers know it. A Kansas farmer told a magazine writer:

> I just wish you could tell us how to get smaller. I'm from the generation that took to farming because of the life-style, and if we made a living that was okay. The best time I ever had farming was when I used horses and came home every evening at five to help Eileen in the vegetable garden. But there's no time anymore.[16]

"Small" farmers have to hustle like the business executives in cities whom they have so often ridiculed. But unlike executives and stock-brokers, farmers don't get such a large return on their investment and they must put in long hours of hard work to get it. For all of their six-figure balance sheets, farmers still do the hard, dirty work of agriculture while the rest of agribusiness makes the easy money supplying them, lending to them, and reselling their products. No wonder so many farmers get disgusted and quit farming for other work.

We've got a huge investment; we can't afford to let it sit idle. The building has to be working for us all the time. That means keeping it at capacity all the time.

—Neil Beck, Iowa pig farm manager
quoted in Warren Clark, "Have We
Broken the Hog Cycle?" *Farm
Journal,* October 1976, p. Hog-34

What's Wrong with Bigness and Mechanization?

Trends in the poultry industry provide a good illustration of what happens as animal production systems increase productivity through capital intensification. Because of the demands of expensive equipment, only large companies can afford to enter the business and to expand production; when they do, smaller operators are gradually squeezed out. A poultry industry magazine reported in 1975 that expansion in the egg business "is being done mainly by the large complex-type operations or corporations made up of producers, hatcheries or processors."[17] The magazine reported "very little, if any expansion" of small (30,000-bird), individually owned egg farms.[18]

The broiler industry is so capital intensive that only a handful of giant firms now produce meat chickens. Over 90 percent of broiling chickens are produced by some fifty large agribusiness companies.[19] In Maine alone, traditionally a poultry state, five companies now produce virtually all of the broilers processed in that state. University of Maine poultry specialist Forest V. Muir estimates that twelve individuals own 92 percent of all table-egg-producing birds in the state.[20] And the largest companies frequently merge, further reducing competition in the poultry market.

Breeding and hatching operations are getting fewer and larger as well. Experts predict that by 1985, only four large drug companies will own all breeding and hatchery facilities. This will lead to, among other things, a further narrowing of the commercial chicken gene pool.[21] It is predicted also that by 1990, only thirty large agribusiness companies "will control U.S. egg production."[22]

The trend to more costly production facilities, whether they be animal factories or field equipment, dictates that only the biggest, wealthiest operators with the most expansionist philosophies will control food production. Everyone else, presumably, will end up working for them if and when jobs are available. As poultry industry history indicates, there will be fewer and fewer jobs available as the trend advances because these interests prefer to eliminate labor wherever possible. In the past twenty-five years, for example, broiler production has increased fourfold while labor requirements have been reduced by 65 percent.[23] Poultry industry experts dote on elimination of labor; in 1977, *Broiler Industry* boasted that "automation can halve plant labor in three years."[24]

Pigs Are Next

Pigs have been called "mortgage lifters" by farmers since the nineteenth century for their reliability in bringing in cash to the farm household. Now the independent farmers' last bastion of cash flow from animal production is going the same way as the poultry industry. Independent farmers already see the handwriting on the wall. As one Missouri pig farmer puts it:

> We've lost the chickens, we've lost the turkeys and we've lost cattle feeding as far as the individual farmer is concerned. The only thing that is left is hogs. It's just a matter of time.[25]

Specialization is spreading in the pig industry. As in the poultry industry, production is being broken up as companies take over the various stages of production. "Farrowing corporations," like poultry hatcheries, supply young animals to growers, who feed them to market weight. Some line up client farmers, require them to build finishing facilities to their specifications, help them arrange financing, and provide weekly management advice. Land grant schools are doing their bit to pave the way for specialization and corporate intrusion into pig farming. They have already made pig farming complicated; now they offer courses and degrees in factory management to add to the supply of trained help needed by the largest pig operations.

Large agribusiness companies, especially the broiler corporations, are getting into the factory pig business. A Pennsylvania company, Pennfield, began back in 1971 as a feed company. It expanded rapidly into factory production of broilers and eggs and has now built a pig factory "as a test program to demonstrate that we can get the performance we need to move into hogs."[26]

Meanwhile, the trend to factories is making it hard on other producers. A study of Missouri farmers revealed that while some producers were increasing production by building new factory facilities, almost half were cutting back or quitting, and many had no plans to expand. When asked about their reasons, they cited problems associated with the trend toward factory production; these included "diseases, breeding difficulties," and "high capital investments."[27] According to *Successful Farming*, "Basically

"The high capital investment it takes to raise pigs today may force the producers out who aren't ready and able to take on the debt load required to raise hogs in confinement units."—Jerry Peckum, Iowa Production Credit Association loan officer, quoted in "Have We Broken the Hog Cycle?" *Farm Journal,* October 1978, p. Hog-34.

the study emphasized a national trend: more hogs and fewer hog producers." [28]

Can agribusiness corporations take over pig production as they have poultry? It's been thought that these animals need too much care and attention to be trusted to hired help, but corporations, with university help, are trying to change that thinking. The factory systems are nearly perfected and the skilled help to run them is in training. Two agricultural experts have been following the trend toward factory specialization and expansion for the past several years. One of their studies concluded:

> We have the feeling that the technology, organization and managerial capacity is now available and in place to slowly but surely change drastically the structure of the hog business. It may well be true that someone can be found "to sit up with the corporate sow." [29]

Like the now-extinct small dairy and chicken farmers before them who relied on small-scale animal production for cash income, many of these pig farmers will be lost to agriculture. As history has shown, when small, diversified farmers lose their cash flow from animal production, they lose their farms as well.

"Farmers" vs. Consumers and Farmers

The broiler, egg, dairy, and beef industries have all reached the level of capital concentration and monopolism that gives them the market muscle to influence prices through restraints on competition. Forty-two of the broiler industry's largest firms, under the canopy of their National Broiler Marketing Association, were accused of restraining competition in a recent suit brought by various supermarkets, distributors, wholesalers, hotels, fast-food outlets, and states that bought their chickens. All but three of the firms have agreed to a settlement under which they will pay out at least $32 million in damages to the plaintiffs, but they admit no guilt to charges that they conspired to fix, maintain, and stabilize broiler prices.[30] Scarcely a year after the settlement was reached, the NBMA members voted to dissolve the organization.

The largest egg-producing firms have formed an organization called EGGMAR to help centralize marketing, to "prevent price erosion," and to bring "more price stability" to egg marketing; as a producer "cooperative," it is legally exempt from the antitrust laws.[31]

Iowa Beef Processors, Inc., one of its subsidiaries, and six of the largest western cattle feedlots were charged with monopolism recently after they made an agreement that would have given them control of feeding, slaughter, and marketing of one quarter of the cattle fed in the states of Idaho, Washington, Oregon, and Montana over the next five years.[32]

But monopolism is legal if it's done right. Like the egg industry, the milk industry takes advantage of loopholes in the antitrust laws designed

Under the layer cages.

to protect "farmer" cooperatives. Associated Milk Producers, Inc. (AMPI), and Land O' Lakes, for example, are high on the Fortune 500 list of the nation's wealthiest corporations, but they are "farmer cooperatives." The giant cooperatives form even bigger federations that monopolize market areas. In 1971, AMPI controlled more than 70 percent of the market in fourteen midwestern market areas designated by the U.S. Department of Agriculture.[33] At present, three federations control 80 percent of the New England milk supply, and one federation controls 70 percent of New York–New Jersey milk.[34] This situation inhibits competitive milk pricing and keeps milk prices to consumers at about 20 cents a gallon higher than the competitive price would otherwise be.

Laws favoring cooperatives are supposed to give farmers a voice and some muscle in dealing with distributors, but these laws are exploited to the hilt by agribusiness for profit at the public's expense. Co-ops are not so good to farmers, either. Their big-business tactics of merger, expansion, and vertical integration have created costly, centralized bureaucracies that reduce farmer control as well as farmers' net share of co-op income. Agricultural economist Truman Graff of the University of Wisconsin has estimated that only about 51 cents out of every dollar spent on milk ends up in the pockets of co-op member dairy farmers; the rest goes to advertising, lobbying, salaries of executives and staff, and other operating expenses.[35] The big co-ops deny membership and marketing benefits to small, independent dairy farmers because they don't turn out a large enough volume of milk. Thus the big factory dairies have an enormous competitive edge over small producers—so much so that the latter are now nearly extinct.

The term *farmer cooperative* has a nice, democratic, American ring to it and the concept could be good both for working farmers and consumers. But, as with so many other labels, the reality behind it is somewhat different. In trying to compete with market-dominating corporate distributors and wholesalers, cooperatives have turned to professional management for expertise, and corporatism has rubbed off. Management interests now overshadow farmer interests. A strict profit orientation prevails. Farmer-members, like stockholders in a corporation, play only a small role in cooperative decision making. Often the co-op management asks corporate executives to sit on the board, invites food firms and conglomerates into partnerships, and even solicits agribusiness corporation membership. Farmer-members have let co-ops slip from their control.[36]

The Hard Sell: Moving Factory Products

An array of industry organizations now push up demand for animal products with slick magazine and television advertising and other promotional techniques. Familiar ones are the American Egg Board's "Incredible Edible Egg" and the American Dairy Association's "Milk Is a Natural" television spots. These groups are funded by farmers from a cut of their checks when they sell eggs, milk, or animals. They are supposed to be producers' organizations, but, like farmers' cooperatives, they benefit and are led primarily by the largest operators. In 1976, for example, when beef industry leaders pushed for a "check-off" scheme (percentage contributions taken from farmers' sales to provide promotional funds), they failed to get the necessary two-thirds vote because of opposition from small and medium-sized cattle farmers. On the heels of that attempt, a new "Beeferendum" was drawn up and the vote required for approval was reduced to a simple majority.[37]

If approved by producers, the beef industry's check-off will bring in an estimated $30 million to $40 million each year.[38] The money will be spent according to the wishes of a "Beef Research and Information Board" to be made up from the ranks of key cattle producers. The whole scheme was set up by an act of Congress; the U.S. Department of Agriculture will appoint the board members and will supervise the vote. If the "Beeferendum" passes, USDA will administer the program and our U.S. District Courts will enforce the laws establishing it. This looks, sounds, and smells like something official and in the public interest. But it isn't. According to one consumer advocate, its proponents have fought attempts to have consumer representatives on the board. If it passes, and is administered like the other "check-offs," half or more of the money raised will go for "consumer education" (advertising of beef), the "information" component of the program.

The other animal products industries also use check-offs to raise money for "research" and "information." With names like National Egg Board, National Council on Egg Nutrition, and National Dairy Council, these organizations bolster the persuasiveness of their campaigns by conveying to the public an image of an entity that is official, authoritative, and a servant of the public interest. The National Egg Board, with the American Egg Board, estimated it would take in $6.3 million in 1979 and over half of it would go for radio and television commercials and other advertising.[39] The dairy industry's United Dairy Industry Association, which includes co-ops and dairy promotional organizations, has some $28 million a year to spend on "research" and other promotion.[40]

The biggest agribusiness corporations, of course, spend additional millions advertising their own brands of beef, chicken, milk, and the other animal products. Here again, the consumer pays a premium for the heavily advertised brands, getting little more than a vague psychological satisfaction in being a consumer of the "best." For all of Frank Perdue's boasts in advertisements about his chickens, it appears from the Consumers Union survey mentioned in chapter 4 that his chickens are, after all,

At last the special day has come.
She is so very proud
As she looks down at her very
    first egg
She clucks and clucks so loud.

It is usually only a few days after she is in the laying house that she lays her first egg. Chickens do actually 'cluck' or 'sing' after laying an egg ... it really seems to make them happy. Incidently, there are no roosters (male birds) in these laying houses. The hen just lays eggs naturally. The rooster is required for a fertile or hatching egg.

The cow now goes with many others
To the pasture to drink water and eat grass.
Some may stop at the salt box
To take a lick of salt as they pass.

In the summer the cows are turned out to pasture to eat grass. The dairy farmer keeps a salt box in the pasture because cows need salt. Cows also need a lot of water. It is important in helping them to digest food and to make milk. Water also helps to keep the cow cool in the summer. She may drink as much as twenty gallons of water a day.

"Educational" coloring books for children, described as "factual story approved by The American Egg Board" and "factual story reviewed by The National Dairy Council and Milk Industry Foundation." Copyright © 1975 and © 1976, Know-about Publications, Inc., Harrisburg, Pa.

no different from any others. But advertisements for branded and trademarked animal products bring profitable results, according to the president of a large meat-packing firm:

> The evidence leads to the consumer message. That's the future . . . the obvious direction. Add value to the products, give them a trademark. . . . Oscar Mayer has led the way. . . . They get 20¢ more for bacon than anyone else.[41]

Animal industries are adopting other kinds of information control in a reaction to attacks against animal agriculture for its wastefulness and for the health risks associated with its products. *Confinement* magazine indicates this defensiveness, claiming that "foods of animal origin are under continuing attack in this country by a broad army of uninformed physicians, misinformed media people and various health-food nuts."[42]

Animal industries appear to be worried that younger children may not have sufficiently robust appetites for meat, milk, and eggs should they wonder, as they are prone to, about where these products come from. To head off such anxieties, they are sending pleasing messages to children about animal production methods. The dairy, egg, and pork industries sponsor coloring books and other materials for children that show farm animals in anthropomorphic, comic book fantasies. There is no hint, of course, of feed additives, stress, crowding, or debeaking.

The Care and Feeding of Experts

Animal agribusiness controls not only demand and markets, but the course of agricultural technology as well. Here its job is not so difficult; all it must do is provide money for the right research. Research on capital-intensive factory methods and systems is naturally high on its list of proper research subjects.

Sometimes agribusiness doesn't have to donate money; its products do just as well. At the University of Illinois, for example, where scientists are working to perfect sheep factories:

> The high-profile aluminum floors were furnished by the Aluminum Company of America. The ¾-inch #9 Saf-T-Mesh was furnished by Wheeling Corrugating Company. The stainless steel plank was furnished by Behlen Manufacturing Company. The low-profile aluminum floor was furnished by Danforth Agri-Resources, Inc.[43]

Once the design is perfected, the firms can realistically expect that their products will go into the manufacture of these factory systems.

Agribusiness firms also use the personal approach to mark the proper subjects for research. Many scientists, like all professionals whose careers thrive on reputation and ego gratification, respond well to recognition and

Experimental sheep confinement unit, Ohio State University, Wooster, Ohio.

back-patting. Knowing this, agribusiness interests dole out scrolls, plaques, titles, and cash awards to scientists for their work. By watching who gets the awards, we can tell which lines of research are deemed worthy. For instance, the scientist who developed DES for use as a growth promotant in feed has won a string of awards, distinguished professorships, and honorary fellowships; he is credited with having "brought science to what had been an 'art'—feeding cattle."[44] Agricultural manufacturers and suppliers like Ralston-Purina, Merck, Pfizer, Shell Oil, and Upjohn, trade associations like The American Feed Manufacturers Association, farm magazines, and industry promotional groups are among the award givers.

Agricultural support industries can rely on a few scientists and quasi-professional organizations to provide expert advocacy when the dangers from pesticides, antibiotics, and other products are exposed. Foremost among these is the Council for Agricultural Science and Technology (CAST). CAST is, indeed, agro-industry's "truth squad":

> The attacks on use of food additives and pesticides have been occurring for years. To counter the emotionalism toward agriculture and to be a spokesman for scientific truth . . . [CAST] was formed in 1973.
>
> It consists of 22 scientific societies, composed of 60,000 agriculturally oriented scientists, making it possible to amass scientific truth quickly and getting it to the right place to dispel developing problems.[45]

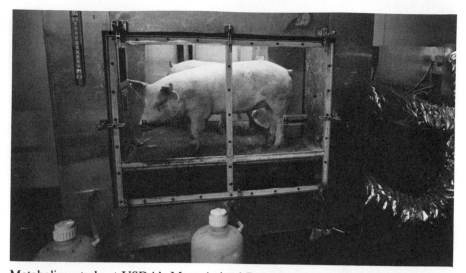

Metabolism study at USDA's Meat Animal Research Center, Clay Center, Neb. "There's no question that drugs and growth promotants are 'where the action is' so far as current research is concerned."—Carey Quarles, Colorado University animal scientist, quoted in "A Look at New Drugs and Future Trends," *Broiler Industry,* January 1977, p. 24.

CAST's annual budget of about $200,000 comes not only from its member societies but also from agricultural trade associations and manufacturers of fertilizer, drugs, and pesticides—including American Cyanamid, CIBA–GEIGY, Dow Chemical, Eli Lilly, Mobil Chemical, Monsanto, Shell Chemical, and Stauffer Chemical. In its seven-year history, CAST has published more than seventy reports on agricultural matters; the majority have tended to downplay the possible dangers involved in modern agricultural practices. Most recently, CAST's pro-industry bias was demonstrated in a dispute with six scientists it had hired to study and report on the use of antibiotics in animal feeds. The scientists' report stated that the practice increases bacterial resistance to antibiotics; CAST reworded their statement in its summary to indicate that the use of antibiotics "might" have that effect. The six scientists quit the study panel in protest. One reportedly believes that the whole purpose of CAST's study project was to produce a document that would be used to counter the Food and Drug Administration's case for a ban on the use of antibiotics in animal feed.[46]

The real parties-in-interest, then, behind present capital-intensive factory farming are the companies that do business in all of the systems, supplies, and products it requires. For decades now, these special interests have influenced research, technology, and opinion—both public and expert—in their own interests rather than in the interests of all farmers or of the public as a whole.

Ethics

The Moral Cost
of
Animal Factories

But I am a heavy eater of beef and I believe it does harm to my wit.
—William Shakespeare, *Twelfth Night,* act 1, sc 3

In 1968 the amount of humanly edible protein fed to American livestock and not returned for human consumption approached the whole world's protein deficit!
—Frances Moore Lappé, *Diet for a Small Planet,* rev. ed. (New York: Ballantine Books, 1975), p. 3

Cruelty is acknowledged only where profitability ceases.
—Ruth Harrison, *Animal Machines* (London: Vincent Stuart, Ltd., 1964), p. 3

THANKS TO WESTERN DIETARY HABITS, nutritional misconceptions, and, more recently, aggressive promotional campaigns, the average American consumes some 607 pounds of animal products each year.[1] Assuming that the nutritional standards set by the National Academy of Sciences and the National Research Council are valid, this is about twice the amount needed for adequate protein nutrition. Many nutritionists believe that the standards themselves overestimate the amount of protein we really need. This overconsumption is not only wasteful; it is harmful to human and ecological health as well. Of course, the animal industries' real concern is expansion of profits. For them, more consumption is better.

Why Do Americans Eat So Much Meat?

In the nineteenth century, the westward expansion of the United States opened up vast new lands that attracted millions of people from hungry,

113

land-scarce Europe. Huge herds of cattle grew fat cheaply, grazing on grassy plains. Cowboys and cattle drives brought meat-on-the-hoof to railheads in Kansas, Missouri, and Nebraska to feed the demand for meat in America's growing cities. At the same time, American society was still largely rural. Land and forage were cheap and a few farm animals provided a reliable and plentiful source of food and some cash income for the farm household; it was not difficult to develop and satisfy an appetite for meat. Waves of immigrants took readily to the new country's meat-based diet, a diet that had been beyond their means in the Old Country. By the turn of the century, Americans were eating about 65 pounds of beef, 65 pounds of pork, and 16 pounds of poultry per person each year.[2]

Soon thereafter, grasslands, woodlands, and pastures fell victim to the need to turn out the numbers of cattle, hogs, and poultry required to feed the growing American population. Prairies were plowed under and forests cleared to grow grain and fodder to feed expanding populations of these animals. As more and more land was converted to raising animals, environmental conflicts arose. The diverse biocommunities of prairies and woodlands lost their habitats to monocultures of corn, wheat, alfalfa, soybeans, or other crops; some species became extinct, others intruded into new areas where they caused dislocations, and still others remained to plague farmers with predations on their crops and livestock. The last would face diminishing chances of success against state and local forces that rallied against them. States established bounties on predatory animals and local communities organized hunts and trapping campaigns to clear the land of crows, rabbits, groundhogs, hawks, and other "varmints."

Science + Business = The American Way

Early in the twentieth century, the science of nutrition began when vitamins and other nutritional elements were identified and their role in health discovered. In the 1920s, the meat-packing industry, with USDA cooperation, began its meat promotional campaign in the public schools, touting their product as practically synonymous with protein, sound nutrition, and good health. The government's involvement in pushing animal products stemmed from an agricultural policy that promoted animal production as a means of increasing farm cash income.[3]

Then, in the 1930s, a tragic form of malnutrition was observed in African children. It was called by its African name, kwashiorkor, and it was discovered that protein deficiency was a major cause. Soon, wherever malnutrition was spotted, nutritionists painted it with the brush of protein deficiency. Actually, much of the malnutrition was simply marasmus, or a plain lack of food and calories. USDA put out charts and other materials

I believe it's completely feasible to specifically design an animal for hamburger.

—Bob Rust, Iowa State University
meat specialist quoted in Gary
Vincent, "Hamburger Cattle,"
Successful Farming, October 1977, p. B-15

that advised consumers to choose from among twelve food groups. Three of these were animal products: dairy products, eggs, and "meat, poultry, and fish." Under intense pressure from livestock interests, these became the "Basic Seven" food groups during World War II. Under continuing pressure from meat and dairy interests, the seven food groups were reduced to the "Four Food Groups" in 1956. Meat and dairy products constitute two of the four, with the implication that these foods should make up half of our diets.[4]

The emphasis on animal protein in diet should have been a passing fad once nutritionists learned better, but the industrial interests concerned saw to it that it stuck. Until very recently, most nutritionists would not stray from dogma on the necessity of animal protein. Their doctrinaire approach, fanned by industry promotional propaganda, led to a steady rise in consumption of animal products as people became more affluent after World War II. Not to eat meat began to appear downright un-American. In 1946, a book entitled *Meat Three Times a Day* proclaimed:

> To argue [that people should substitute grains for animal protein] is to misunderstand the spirit of Americans and what lies back of our country's greatness and productivity. Instead of talking about how low our meat consumption can be cut . . . we should be working at increasing it to a pound a day or even more.[5]

This exaggerated role of animal protein in diet has had tragic consequences. Many poor people have inadequate diets because they are led to spend a disproportionate amount on expensive animal protein. It is estimated that about a million Americans rely on pet food for a significant part of their diet, instead of buying cheaper, more wholesome beans, grains, nuts, and other foods.[6] For their part, the more affluent have had their lives shortened by a diet too high in protein and animal fats and too low in roughage.

THE FOODS YOU NEED EVERY DAY

MEAT, POULTRY OR FISH
One or more servings
Eat any kind of meat (beef, veal, pork or lamb). Include variety meats often, such as liver, kidney and heart.

MILK
Adults 1 pint—Children more
Drink milk or eat cheese and foods prepared with milk, such as custards, creamed dishes, soups and ice cream.

EGGS
One (at least three a week)
Have it cooked any way desired or in combination with other foods.

POTATOES
One or more servings
Choose either white or sweet potatoes prepared in any of a variety of ways.

VEGETABLES
Two or more servings
Eat green and yellow vegetables often. Include salads or other raw vegetables.

FRUITS
Two or more servings
Eat all kinds of fruit. Have oranges, grapefruit, tomatoes or berries often.

BREAD AND CEREALS
As needed
Select enriched breads, cereals, etc., or those made of whole grain products.

FATS AND SWEETS
As needed
Include some fat. Enjoy sweets after other foods needed have been eaten.

The "Basic Seven," circa 1940. The eighth group here, fats and sweets, was considered an "extra" because those foods contained few nutrients relative to calories.

Heresy and Subversion

The assault on the meat-equals-nutrition citadel began in about 1970 with the publication of Frances Moore Lappé's brilliant book *Diet for a Small Planet*. Though criticized since by some progressive nutritionists for its overcomplicated approach to plant protein "complementarity" (the mixing of certain kinds of plant protein in the same meal), her book exposed "the incredible level of protein waste built into the American meat-centered diet."[7] She raised the ethical issue posed by one country's addiction to the luxury of animal protein while much of the rest of the world starves. Lappé explained how American demand for meat causes half our cropland and upward of 90 percent of our corn, oats, barley, sorghum, and soybeans to go to the fattening of our meat animals and how those animals return only one seventh of these grains back to us in edible meat, thus wasting the other six sevenths of this grain (118 million tons back in 1971).[8] In addition, the rich Western countries' demand for meat results in plant proteins becoming too dear for people in the poorest, hungriest countries. Lappé urged those concerned to change to meatless diets, to eat lower on the food chain, and to take other steps in the direction of a rational use of our world's agricultural resources.

Lappé's book sold over a million copies and raised a controversy that still lives. Just as Rachel Carson's *Silent Spring* lifted awareness of environmental pollution, *Diet for a Small Planet* opened eyes to the relationships among Western diets, world hunger, and environmental despoliation. In the wake of the book, newspapers and magazines ran features about the controversy and the new dietary awareness. The meat industry got into a dither over all of this "antimeat propaganda" as many people moved toward vegetarian diets and eating less meat. In this changing climate of dietary consciousness, a jump in meat prices in 1973 led to consumer boycotts, and per capita meat consumption fell back to 1967 levels.

Meat industry leaders may have considered Lappé's book harmless heresy, but the drop in consumption was not good for business. They saw to it that the downturn in meat consumption did not last long. The propaganda organs began rolling to nip America's sudden new guilt over meat eating in the bud. If we cut back on meat eating, they argued, farmers would cut back on grain production and there would be less food for everybody. This was more threat than reasoning; it assumes that cropland and grain freed by reducing animal consumption would never be diverted into human food production. Meat advocates claimed that animals produce food for humans from forage inedible by humans, so that land too poor to support human food crops can produce food in the form of meat. Regardless of that fact, however, about 70 percent of all beef cattle are fed humanly edible grain and other concentrates in the finishing stage. Moreover, this argument ignores the wasteful practice of grain-feeding poultry and pigs. Because these animals are unable to digest forage, roughage, or plants from poor cropland, they compete directly with humans for food. Counterattack propaganda to the contrary, livestock still consume about three quarters of our grain crop and take up cropland that could either support other plant and animal life or provide humans with a sensible and more varied diet.

Meat Revival Evangelism

In 1975, a record corn crop lowered grain prices and cattle started going back into the feedlots. Meat eaters, tired of the affronts to their moral integrity, began to fight back and to call for a return to the old-time steak religion. The politics of protein took a sharp swing to the right and the meatless diet came to be considered a subversive activity. Soon, meat consumption per capita and meat industry gross profits climbed to all-time highs.

Even so, the meat industry is not out of the woods. There has been mounting evidence linking excessive consumption of animal products with heart disease, stroke, cancer, diabetes, gout, osteoporosis, and a host of infectious diseases. Much of this evidence was presented before the Senate Select Committee on Nutrition and Human Needs in hearings during 1976. In January 1977, the Committee announced its "Dietary Goals for the United States," containing recommendations that Americans decrease consumption of meat, heavy fats, whole milk, butterfat, and eggs. Before the ink dried, the Meat Board and the American National Cattlemen's Association were stirring up pressure to have the report withdrawn on the grounds that there was "insufficient scientific evidence" to support the recommendations. They managed to force the committee back to the hearing rooms in March 1977 to take down "supplemental evidence" on the dietary question. Later that year, the second edition of "Dietary Goals" appeared—minus the "eat less meat" language but otherwise substantially the same report as before. Then, in September 1979, the Surgeon General of the United States issued a report entitled *Healthy People,* which recommended that people eat less meat

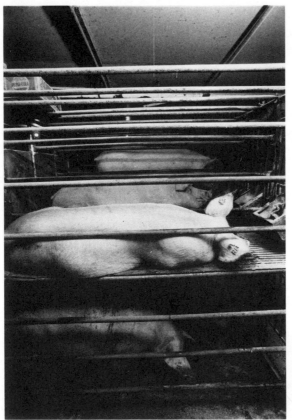

"They can still eat—total darkness has no effect on their appetites."—Harry Sterkel, Jr., "Cut Light and Clamp Down on Tail Biting," *Farm Journal,* March 1976, p. Hog-6.

The 37 million elementary and 15 million high school students in the United States constitute a special [National Livestock and] Meat Board audience.
—*Meat Board Report, 1974–1975*
(Chicago: National Livestock and
Meat Board, 1975), p. 23

and make other dietary changes along the lines of the original "Dietary Goals." On its heels came another report, from the American Society for Clinical Nutrition, echoing the conclusions of both of these reports (one of the panelists recommended that meat be used only as a condiment).

The meat industry may hope the worst is over, but it's not taking any chances. The great war over a sensible diet is on. As part of its strategy, industry generates myths about protein and stimulates appetites for more and more meat; at the same time, it gestures to "rising expectations" and the "need to supply superior meat protein" to an ever-widening circle.

It would be naive to expect long-range objectivity about diet, nutrition, and world resources where short-run profits are so easily made. The truth is that the carefully fostered American meat addiction creates market conditions that ensure the wasting of grain on livestock. Farmers simply make more money on corn in the form of flesh. According to *Beef* magazine:

> The corn you feed your hogs during January, February and March will be bringing you about $3.90 a bushel! That's almost double what you'd get for it if you were to sell it as cash corn rather than as pork.[9]

Meanwhile, the misapplication of technology and the misappropriation of resources go on. Because of exorbitant animal production, each Westerner uses up to six times more grain per day than a citizen of the hungry world. While some 460 million people (10 percent of the world's population) suffer *severe* malnutrition, American farmers shoot calves, let cattle starve in barren fields, and dump milk to protest against low prices.[10] Less dramatic, perhaps, but no less arrogant is the practice of pouring whey, dry milk, fish meal, oil seeds, and other materials potentially nutritious to humans into our livestock feeds to produce hefty animals quickly.

It is true, as the livestock industries are quick to point out, that under present economic conditions there is no guarantee that if grain and resources were diverted from our livestock, they would be distributed to the world's hungry. Changes more far-reaching than personal dietary preferences in affluent countries must be effected if world food supplies are to be shared more fairly. But there is a danger that our wasteful habits will spread. American agribusiness is working feverishly to spread them. Industry promotional organizations are working with the USDA to expand exports of meat and animal products. Farming magazines like *Farm Journal* praise the efforts of developing countries in "climbing the protein ladder":

> In fact, enlarging and diversifying their meat supply appears to be a first step for *every* developing country. They all start by putting in modern broiler and egg production facilities—the fastest and cheapest way to produce nonplant protein. Then, as rapidly as their economics permit, they climb "the protein ladder" to pork, to milk and dairy products, to grass-fed beef, and finally, if they can, to grain-finished beef.[11]

Russia, Egypt, China, and many other countries are expanding their animal populations, increasing the grain drain, and placing an even greater burden on the environment. American agribusiness support companies are only too happy to send their consultants and salespeople to these countries to help them set up American-style animal factories. One has to wonder, and worry about, where all of the grain and resources to keep these factories running will come from. American agribusiness knows. It's all part of the plan to build markets for the grain, drugs, hardware, and animal "seedstock" provided by the same companies and folks who brought *us* factory farming.[12]

Can we afford to let them get away with it? With the world population now at 4 billion and expected to double by the end of this century, can we afford to waste cropland and destroy food by running it through animal factories, whether here or elsewhere? Will we realize too late that the entire world cannot "climb the protein ladder" to America's wasteful dietary standards? The tragic myths of the "superiority" of animal protein and the "efficiency" of its production in animal factories must be destroyed before it is too late.

Down (and Out) on the Farm

Among the less visible, less tangible costs to society from overreliance on factory farm technology is the disruptive impact this mode of farming has on rural communities. Consider what has already happened in the poultry industry. As late as 1959, nearly 60 percent of all broilers were grown by independent farmers and sold to processors on the open market.[13] Now virtually all broilers are produced by "contract growers," most of them former independent farmers who agree to provide space and labor to grow company birds on company feed, following company specifications. Farmers who can get the loans to put in the required

The end of a way of life.

USDA photo

buildings and equipment find that they don't make much money for the huge debt load and long hours of labor that are the consequences of their new company affiliation. And many independent farmers are squeezed out of business by their inability to get the necessary capital. Once farmers become "poultry peons" tied to big companies, they must keep the chickens flowing—even when losing money—just to meet payments on their mortgaged farms and homes.[14]

If the trend in pig farming continues to follow the line of development established by the poultry industry, pig farmers and the communities they live in may be headed for the same fate. One study of the pig industry concluded that if big companies gained control of the whole business, from selling feeds to marketing the meat, "producers would lose 5 to 10%

in net incomes, rural communities up to 12% in level of economic activity. And consumers would pay up to 12% more for pork."[15]

The decline of small farms has caused drastic dislocations of society, from the rural areas directly affected all the way to the urban core. Rural communities lose people and vitality, while urban areas see increased crowding, welfare burdens, and unemployment as the expropriated farmers seek new livelihoods. At the beginning of World War II, there were 6 million farms and a farm labor force of nearly 11 million in the United States; today, there are 2.7 million farms and about 4 million farm workers.[16] Much of this shift in population was caused by political and economic policies that made life difficult for small, labor-intensive farms, particularly the policy of mechanization in agricultural production. In what has been commonly described as "economic cannibalism from within agriculture," small farms were replaced by large farms that could afford the machines and new production methods. Small, self-sufficient farms, with the families who lived on them and the diverse crops and animals they raised, were phased out. "Subsistence" agriculture—"living off the land" or farming as a way of life—did not fit into the policy makers' plans and was replaced by agribusiness.

The land was too valuable for subsistence farming. Human labor is more productive in the service of a corporation's goals. Subsistence farms were seen as a drain on the economy and their occupants were pushed out to bloat the cheap labor pools around the urban areas of the industrial North.

Despite all the pressures toward size and specialization, a few people still realize the value of their small-scale, family-run, diversified farms, and keep them productive and profitable without the massive inputs of capital and energy used on agribusiness farms. But agriculture experts perceive these farms as aberrations from the norm. When an article appeared in *Farm Journal* extolling the virtues of the "natural tide of history" (the trend toward large, energy-intensive farms), several readers protested.[17] The big farms, they complained, are able to get the good land that could make the difference between success and failure for smaller farmers. One reader wrote: "I hope that sometime, such leaders of agriculture as *Farm Journal* will wake up to a consideration of the social and ethical factors involved."[18]

Animals Are Not Machines

There is a final ethical cost to the animal factories. We often seem to assume that animals were put here for us to use as we please. But no good reason can be given for regarding animals as things. They are not things. They can feel pain. They can suffer frustration and boredom. They have lives of their own to lead. Much of the "progress" in factory farming

"We consider that calves should have sufficient room to be able at all ages to turn around, to groom themselves, and to move without discomfort."—*Report of the Technical Committee to Enquire into the Welfare of Animals Kept under Intensive Livestock Husbandry Systems,* Command Paper 2836 (London: Her Majesty's Stationery Office, 1965), paras. 147–49.

methods raises an ethical question: do we have the right to make animals live miserable lives, just to satisfy our taste for a diet so rich in animal products that it exceeds any sane nutritional requirement?

In a letter to *Farmer and Stockbreeder,* an English farm magazine, a Coventry veterinarian wrote this response to a report on then-new cage systems for pigs:

> May I dissociate myself completely from any implication that this is a tolerable form of husbandry? I hope many of my colleagues will join me in saying that we are already tolerating systems of husbandry which, to say the least of it, are downright cruel. . . . Cost effectiveness and conversion ratios are all very well in a robot state; but if this is the future, then the sooner I give up both farming and farm veterinary work the better.[19]

"Ten years of confinement raises more questions than answers."—Dale McKee, Rio, Ill., pig farmer quoted in *Hog Farm Management,* March 1979, p. 124.

Fortunately, these views are shared by some professionals in the United States. Shortly after *Confinement,* the omnibus factory farming magazine, came out, a retired farm veterinarian sent in a thoughtful letter. In part it read:

> More and more, I find myself developing an aversion to the snowballing trend toward total confinement of livestock. . . . If we regard this unnatural environment as acceptable, what does it portend for mankind itself? . . . How can a truly human being impose conditions on lower animals that he would not be willing to impose on himself? . . . Freedom of movement and expression should not be the exclusive domain of man. . . .
>
> What [then] of human behavior in [the future]? Will it sink to the nadir of contempt for all that is naturally bright and beautiful? Will all of us become tailbiters without recognizing what we have become? [20]

Animal factory technology seeks to increase productivity and efficiency in food accumulation. But when sows are routinely dosed with hormones and surgically scraped to extract greater productivity and efficiency in protein production, doesn't one wonder whether this end might not better be achieved in another way? When calves are forced into an anemic, neurotic condition just to satisfy the gourmet's desire for pale flesh, increased productivity is not even the goal. Productivity for and catering to the whims of the market may be all right in the plastics or automobile industries, but it can be cruel and abusive when the factory method is applied to animals. Moreover, one must challenge any claims of productivity and efficiency in a mode of agriculture that has inherent in it so great a waste of land and resources. As we have seen, there is no real human need being satisfied by these methods, just habits and appetites. The justification of necessity is frequently offered when our dealings with other animals are concerned; but the "necessity" so often turns out to be entirely unnecessary that one suspects it to be a tacit admission of our awareness that those dealings are morally indefensible.

Animal factories are one more sign of the extent to which our technological capacities have advanced faster than our ethics. We plow under habitats of other animals to grow hybrid corn that fattens our genetically engineered animals for slaughter. We make free species

"In principle we disapprove of a degree of confinement of an animal which necessarily frustrates most of the major activities which make up its natural behavior. . . . An animal should at least have sufficient freedom of movement to be able without difficulty to turn around, groom itself, get up, lie down and stretch its limbs." —*Report of the Technical Committee to Enquire into the Welfare of Animals,* para. 37.

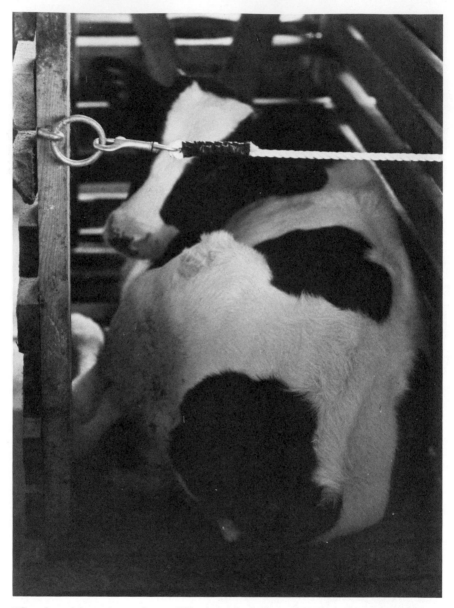

"Good veal has always been difficult to find. But recently a Dutch process has come to our shores and is giving us a limited quantity of much finer veal than was generally available before. . . . The process consists simply of taking calves from their mothers' milk to small stalls, where they are fed with vitamins and powdered milk that contains no iron to darken the flesh. Also, the calves are kept comparatively quiet during their milk regime. Thus, they have delicate whitish-pink flesh and clear fat and are deliciously tender."—James Beard, *American Cookery* (Boston: Little, Brown & Co., 1972), pp. 331–32.

extinct and domestic species into biomachines. We build cruelty into our diet.

Our failure to accept animals as beings entitled to ethical consideration in their own right is a barrier to any genuine sensibility in our relations with nature. We lose much that is dear in the process. As two writers who took a look recently at poultry factories concluded:

> We would insist that economics aside, the story [of the industrialization of chickens] has a much deeper moral: Unrestrained technology, fueled by the desire for larger and larger profits, exacts a price in terms of human values that we can no longer afford to pay. There are invisible costs as well as visible ones in the destruction of chickens.[21]

There are many, many costs in our present methods and systems of food production. Agribusiness experts would have us know only the benefits. They use "cost-benefit" analyses to justify the use of antibiotics in feed or chemical growth promotants, or nitrites to cure meats. They assert that the benefits to consumers from these uses outweighs the risks involved. But if this sort of test is to have any validity in agricultural affairs, it must take *all* the costs of factory methods into account: costs to the health of consumers who dine on fatty, chemically dosed, antibiotic-fed animals that never exercise or see sunlight; costs to the environment from the accumulation of huge quantities of noxious animal wastes; costs to our limited stores of fossil fuels; costs to the starving, whose lives might be saved by the food we are wasting; costs to the land, which is forced to produce more and more grain to be turned into meat; costs to wildlife, whose habitat is destroyed to grow grain; costs to the quality of life for small farmers who, getting no support from the Department of Agriculture or the agricultural research establishment, can no longer compete with big business and must leave the land; costs to the animals themselves, confined, crowded, bored, frustrated, and deprived of most of their natural pleasures; and finally, costs to our own self-respect.

We do not pretend to be able to quantify these costs. That problem we leave to the agricultural experts who are so fond of cost-benefit analyses. But we are confident that if adequate account is taken of the costs of the ruthless application of technology to animal rearing, they will decisively outweigh any benefits we obtain from animal factories.

ANIMAL FACTORIES

Toward a Better Way of Life for Consumers, Farmers, and Farm Animals

Food is a cultural, not a technological, product. A culture is not a collection of relics or ornaments, but a practical necessity, and its destruction invokes calamity. A healthy culture is a communal order of memory, insight, value, and aspiration. It would reveal the human necessities and the human limits. It would clarify our inescapable bonds to the earth and to each other.
—Wendell Berry, "Land Reform Starts on the Land and in the Heart," *Environment Action Bulletin,* October 19, 1974

FACTORY METHODS OF ANIMAL PRODUCTION are not, as some agriculture experts claim, the inevitable result of a "natural tide of history." They are the product of decades of government policy and corporate profiteering. Although the trend is reversible, the forces behind it are well entrenched. Therefore, there can be no immediate end to factory methods; it will take patient struggle to bring sanity and humanity to farming. In the meantime, farmers and consumers can begin by stopping their own contributions to the progress of factory farming.

Both farmers and consumers must learn more about the vast no-man's-land between them: the systems of manufacturers, carriers, distributors, wholesalers, and others that supply farmers and deliver food to consumers. At present, each side tends to have a short-sighted view of agriculture. As buyers, consumers are concerned primarily with food appearance and pricing at the supermarket. For them, farmers and their problems seem far away; consumers deal with food only as it comes from the store. As sellers, farmers concern themselves with prices at the market for their crops. Too many, unfortunately, have been swayed by agribusiness companies into blaming their problems on "consumerism" and governmental regulation. Neither farmers nor consumers would grumble so much if they understood the more important problems that lie beyond their points of sale or purchase.

The very first obstacle to be eliminated, then, is the phony war between farmers and consumers. As in conflicts between the races or sexes, it is easier to blame or ignore the other side than it is to recognize the pervasive, hidden attitudes and economic forces that make life miserable for everyone. Geographic separation may make it impossible for farmers and consumers to literally join hands, but they can join minds and recognize that they are the most important parties in humans' most fundamental social activity: the conversion of energy and resources into food.

Primarily, agriculture should provide people with a sensible, healthy diet. Our national agricultural policy should promote the kinds of farming that do so, and it should guarantee information, technology, and economic incentives to people who want to run these farms. There are many dedicated, progressive people in consumer, food, and farm organizations and in the U.S. Department of Agriculture making efforts toward this sort of agricultural policy. But they are up against agribusiness interests who see agriculture primarily as a provider of profits and balancer of trade deficits. These interests control agricultural policy and technology, and they want to keep it that way. As the fray intensifies in the next few years, pro-food and farm people will need a lot of popular support. Consumers and farmers alike will be subjected to propaganda and threats of increased food prices from those who dominate farm and food policy. It will be necessary to go to some trouble to get information and analysis on the issues because newspapers and magazines, reliant as they are on food-establishment advertising, are not likely to be consistently objective.

A movement for these changes is already under way and it needs your support. In their recent book *Food First,* Frances Moore Lappé and Joseph Collins wrote:

> On college campuses, in religious organizations, among certain state and national legislators, in co-op movements, and among ecology and natural food groups there is a feeling that food is the right place to start to focus attention and energy for change. . . . This is definitely not a movement only of young people. It includes farmers' groups that have been fighting many of the battles we have just discovered for decades. Now they have new allies.[1]

Let's look now at some specific steps that can be taken by consumers and farmers to undermine the present trend toward factory farming.

What Consumers Can Do

The first priority for the consumer is to question the alleged necessity of animal products in diet. Chances are you eat far more of these products than are recommended even by present meat-oriented standards, because: (1) you have been hooked on the protein myth, and (2) you have acquired tastes for them that have made them your main foods. Evidence now emerging indicates that this overconsumption of animal products does more harm than good. Excess protein is not stored for a lean day as are fats and carbohydrates; it must be turned over, "metabolized," by the body and eliminated as waste. If you have too much protein in your body, your metabolic rate "idles" faster than normal, shortens the life span of some cells, and contributes to premature aging. Urea, a by-product of protein metabolism, builds up in the bloodstream, burdens the kidneys, and increases the body's demand for water to "flush" the system. High-protein diets have been linked to neurological problems in children because of related effects on body chemistry. They are also believed to upset the body's calcium balance and to contribute to osteoporosis, a bone-thinning disorder that affects millions of elderly persons—especially women. By now it is virtually common knowledge that diets heavy in meat and animal fats have been linked to a higher incidence of heart disease, colon cancer, stroke, and other degenerative diseases.[2]

Because of this evidence and the growing body of opinion that protein needs have been exaggerated, protein is now one of the most controversial elements in our diet. Books listed in the General References at the end of this book offer a more complete discussion of protein nutrition than can be provided here, but a few brief points should clear up most of the myths about nutrients:

1. American consumption of animal protein alone averages twice the amount recommended by the National Academy of Science and the National Research Council. Most of us could simply eliminate meat, poultry, and fish from our diets and still get all the protein we need from the bread, nuts, peas, beans, eggs, and dairy products that we normally eat.[3]

2. The NAS-NRC Recommended Daily Allowances (RDAs) are calculated with a 30 percent safety factor to allow for individual differences or extra needs because of injury or stress. Thus, most of us need about two thirds of the amount of protein that the RDA says we need.

3. The biggest myth of all is that protein from meat is "superior quality"—a designation fostered by the meat industry. Actually, your body needs not protein but eight amino acids that it cannot synthesize. Animal products have been adjudged "superior" because they supply all of these. But so do grains, legumes, nuts, and other plant parts, although

one or more may be "low" in this or that amino acid. Hence the notion that plant protein is "incomplete." This fact makes little difference because people usually eat foods in combinations that supply all the essential amino acids. It's as complicated as a peanut butter sandwich. According to some researchers, the lesser amounts of amino acids in some plant foods may be adequate. These nutritionists look at "nitrogen balance" to assess protein nutritional adequacy. When the balance is positive, protein nutrition is considered adequate. Potatoes, corn, rice, and wheat fed as the only protein source in the diet have kept humans in positive nitrogen balance. Of course, no one recommends such a monotonous diet; it just shows another flaw in the thinking that we *need* "superior" or "complete" animal protein.[4]

The surest way to avoid helping factory farming is to stop consuming its products. You can start by refusing to eat "milk-fed" veal offered in fancy restaurants, and grain-fed beef. Chances are the pork, poultry, eggs, and milk sold in your supermarket come from factory farms, so you should look for them elsewhere. If you feel that you must have these products, cut down on your intake and get them from an "organic," "health," or "natural" food store, or try one of the food co-ops in your area. You may pay higher prices at these places, but if you cut down your consumption you will spend no more overall. You should ask the store or co-op to check up on its farmers and suppliers because, in the absence of inspection and labeling laws, some of them wink at the concern about how food is produced and try to pass along factory-farmed products to get the better prices offered for "natural" or "organic" produce.

You may think that all of these changes are just too much trouble. Maybe you don't have a food co-op or health food store in your area and you don't have time to track down farms where you can buy animal products directly. In such cases, just do what you can. Don't be so rigid about refusing animal products from factory farms that your new food habits become too difficult and you are tempted to drop the whole effort. You will find that in most restaurants there are few main dishes or sandwiches containing no animal products, but the typical menu amounts to a complete list of animal products with a few plant foods listed in fine print as "side dishes" (what better evidence of the animal-protein bias in our dietary habits?). Again, the chances are that these animal products come from factory farms. You will have to make some compromises until more food outlets respond to consumer demand for foods other than factory-farmed meat, milk, and eggs. In the meantime, try to become more aware of your diet and of where your food comes from; learn about nutrition, food, and agriculture.

While we're on the subject of dietary awareness and change, this is perhaps a good place to list the original version of "Dietary Goals for the

United States" prepared by the Senate Select Committee on Nutrition and Human Needs. Before they were watered down under meat industry pressure, the goals suggested the following changes in food selection:

1. Increase consumption of fruits and vegetables and whole grains.
2. Decrease consumption of meat and increase consumption of poultry and fish.
3. Decrease consumption of foods high in fat and partially substitute poly-unsaturated fat for saturated fat.
4. Substitute non-fat milk for whole milk.
5. Decrease consumption of butterfat, eggs and other high cholesterol sources.
6. Decrease consumption of sugar and foods high in sugar content.
7. Decrease consumption of salt and foods high in salt content.[5]

Why not just go vegetarian? A meatless diet offers much more variety than the restrictive rite of flesh eating. There are plenty of good books on vegetarianism that discuss the benefits better than we can here; we have listed a few of them at the end of this book. Try a few vegetarian meals each week for a while and see if you don't find yourself leaning more and more toward these foods and away from heavy, meat-based meals.

Vegetarianism is on the rise all over the country. Even the National Livestock and Meat Board acknowledges that the "slow, steady growth of vegetarianism . . . is not a fad."[6] A Roper poll conducted in October and November 1978 indicates that in the United States there are about 7 million vegetarians and another 37.5 million people who are careful about their meat consumption, while 78 percent of the population acknowledges the merits of the health, economic, ethical, and other reasons for being a vegetarian.[7] Apparently, long overdue changes in our diets are beginning to be considered and pursued by many people. If only we could make agriculture, food companies, restaurants, and the rest of the food establishment deliver the goods!

To aid in the transition to a meatless diet, you may want to try some of the meat substitutes until your new appetites and culinary skill are more developed. *Tofu* (soybean curd) and *miso* (a product of soy and rice fermentation)—dietary staples in the Orient for thousands of years—are now available in many food stores. Several varieties of meatlike "analogs" made from textured soy protein are already available and steadily encroaching on the animal products market. Any of these can be used in place of meat in soups, casseroles, spaghetti sauce, and many other familiar dishes. It is true that some brands are heavily processed with artificial coloring and flavoring, but then so are processed meats. Look around; don't let the few "bad apples" keep you from looking for brands composed of wholesome ingredients.

The processes for extracting protein from soybeans, wheat gluten, and leafy plants are new, but they can supply wholesome, nutritious foods in convenient form without the problems associated with meat production. These processes could eliminate not only the waste of soil and grain, but the contaminants and other undesirable substances found in animal products.[8] And direct conversion from plant materials would eliminate most, if not all, of the expensive disease control programs, pollution problems, and carcass inspections that go along with animal agriculture.

Some people are averse to any form of "processed" food. But there are good processed foods—bread and cheese may be the oldest. With a little artistry and culinary inventiveness, plain old beans, seeds, nuts, leaves, and other plant proteins could be turned into convenient forms with taste appeal and nutritiousness that would please the most reactionary taste-buds. Soybeans and seeds? Ugh! we think. But on the eve of the production of cheese we might have had the same reaction to the suggestion of making food from spoiled milk. Those delicate Bries, mellow Swisses, and robust Cheddars had not yet been tasted.

These changes in food preference should help independent farmers and farm animals as well. Demand for a wider variety of goods should offer farmers more choices in raising crops, less vulnerability to environmental and economic forces, and, perhaps, better financial security. Lower demand for factory-farmed animal products should lower prices, which would make high-overhead factory systems unprofitable. Studies at Iowa State University show that a farmer using a pasture system can market 600 pigs per year with lower production costs and with less than a third of the capital investment involved in a total confinement system.[9] Even agribusiness-oriented farm magazines advise farmers to go back to pasture and quit using expensive drugs and feed additives when farm prices fall. Apparently, present demand for animal products is still so high that the market is "fat" enough to support expensive, capital-intensive production methods. In a tighter market situation, small, independent, labor-intensive animal farms should have a competitive edge over the large agribusiness factories.

At any rate, consumer demand can make a difference. In Europe, some farmers have gone back to traditional systems because of public awareness of factory-farming methods. In the Netherlands, for example, some farmers have switched to free-range systems for chickens because they have learned that consumers prefer their eggs to eggs from cage-reared birds.[10]

Individual dietary changes will not be enough, however. While you get your food shopping, preparation, and eating habits under control, you should work actively toward broader changes in agriculture and food policy. Since consumer demand affects food supply, we should begin making the following demands:

1. Demand a prohibition on the use of antibiotics, growth promotants, and other feed additives in animal agriculture. The Food and Drug Administration's efforts to ban or regulate these drugs are still underway, but they are being stymied by drug and agribusiness corporations. Without these shortcuts to genuine animal care and health, animal losses in crowded factories would be so great that factory systems and methods would not be profitable.

2. Demand an end to the public subsidies that prop up factory farming. If society is to subsidize agriculture, it could make much better choices about the kinds of production to be supported and the kinds of food to be produced.

3. Demand an end to tax-supported research and technological development of factory systems. The present funding scheme is one big boondoggle for drug and equipment manufacturers. Demand that this money and expertise be directed instead to work on farming methods that farmers can afford and manage, and ones that give consumers safe, wholesome food.

4. Demand local markets and food cooperatives where farmers and consumers can trade directly. Every community has a square or park where space could be set aside for outdoor markets. Find the food cooperative in your community; if there is none, start one.

5. Demand meatless meals and nonfactory farm products from restaurants, hotels, airlines, caterers, school lunch programs, and all other public food outlets. Let them know that you are aware of where food comes from and that you are worried about food produced by factory methods.

6. Demand labeling laws that would mark all factory-produced animal products. (Don't settle for a statement to the effect that the farming systems have been approved by an animal welfare organization; there are some that will rubber-stamp anything just to get their name around.)

7. Demand that supermarkets and other food outlets separate factory and nonfactory foods. There is precedent for this in state laws regulating the labeling and display of kosher foods and, in some states, "organic" or chemical-free foods.

8. Demand a tax on meat and animal products that would provide funding to subsidize the production of other crops. This would be no more absurd than our present policy of subsidizing the production of what are essentially luxury foods. If people want to continue to prop up costly, risky animal production, they should have to pay a premium and the premiums could be channeled toward the support of better foods and production methods.

9. Demand an end to meat industry propaganda in local schools; demand to know how nutrition is being taught to your children.

10. Demand a turnaround in U.S. Department of Agriculture policy so that it puts good food and farm livelihood first. The present prevailing

pro-agribusiness bias is a national scandal that has driven millions of farmers from the land and saturated consumers with junk food.

11. Demand land reforms and zoning laws that would restore small, diversified farms closer to populated areas. Too much of the cost of food goes to transportation, handling, and profiteering as food moves from the farm to the consumer.

12. Demand that food products be labeled to carry the name of corporations owning the brand line. This would expose the monopolism behind the myth of a competitive food industry—and the lie that your ham, eggs, milk, etc., come from good old Farmer Jones down on the farm.

13. Demand an end to the animal products industries' "check-offs," which gouge consumers and small farmers for advertising that props up our wasteful diet weighted toward animal products.

What Farmers Can Do

Since the 1920s, government policy has pushed animal production as a means of boosting farm income.[11] It boosted farm income so well that agribusiness firms got into the act and began taking both farms and farm income. Now that the policy and its implications have culminated in factory farming, it's time to start making some basic changes.

In our attempts to eliminate factory farming, we have to be careful not to create new problems. Independent farmers are in a precarious economic position and abrupt, forced changes could ruin many who have invested in factory systems (although, as we shall see, many could easily return to traditional methods and still farm profitably). But we should be wary of an across-the-board substitution of traditional methods for factory methods. Although traditional farms are, in general, more ecologically sound, efficient, and humane than factories, there are gross exceptions. For example, the return of feedlot cattle to pasture and range systems, as advocated by some reformers, would sharply extend the deforestation, overgrazing, desertification, erosion, and other environmental damage now being caused by these livestock production methods.[12] In addition, animals maintained in the open in severe climates often suffer in stormy winters and hot, dry summers. Nor is a return to traditional methods a realistic option for every independent factory farmer. Many factory pig and dairy farmers have too little land either to support their animals on pasture or to bear other crops. As for independent factory poultry farmers, there just aren't many. Even if other farmers were to go into traditional poultry farming, in most areas they would have difficulty finding reliable markets for their relatively small output.

We cannot, then, expect to solve all of the problems discussed in this book simply by going back to traditional animal production practices. If we are to improve diet, environment, and the survival chances of independent farmers, we must reduce our reliance on animal production while promoting diversification in agriculture.

In the meantime, however, farmers can raise animals successfully without using factory methods. There is ample evidence that traditional husbandry methods are productive and profitable. In fact, many farmers who had set up specialized factory facilities have, for purely practical reasons, abandoned them and gone back to more traditional methods. Typical of these is an Illinois farmer who, having too many problems with pig health and equipment maintenance in his total-confinement buildings, went back to small shelters with outdoor pens and runs. There are no furnaces, fans, or antibiotics on his farm now; he believes his pigs do better on fresh air, sunshine, and room to move about.[13]

Despite their bias toward animal factories, farming magazines contain occasional reports of pig farmers who have abandoned or modified strict confinement systems in favor of open pens, shelters, and bedding. In most cases, these farmers gave up on factory facilities because of their greater disease losses, poorer breeding performance, smaller litters, and higher energy costs.[14] In reference to the recent history of factory methods, one Illinois farmer wrote to *Hog Farm Management* that "ten years of confinement raises more questions than it answers."[15] Although the number of factory pig farms in his state had grown substantially from 1963 to 1975, the number of pigs weaned per litter did not change and the amount of feed required to produce 100 pounds of weight gain rose from 409 to 428 pounds in the same period. Moreover, on the farms surveyed, the 1975 pigs required about a month longer to grow to market weight.[16] Some farmers are beginning to question factories and blind expansion, because they are discovering that "if we overcrowd our facilities, we must pay the price of slower gains, higher death losses and poorer feed conversion."[17]

Even a few animal experts are catching on to the false claims about the efficiency of factory systems. Researchers at South Dakota State University found that pigs reared in concrete-floored shelters open to outside pens gained weight faster and more efficiently than pigs in slatted-floored confinement buildings. The reasons for the difference, one of the researchers believes, are that the confined pigs were more crowded than the other group and were annoyed by bad air, noise, and other confinement conditions.[18] Studies at North Carolina State University show that chickens reared in "loose" housing had these advantages over those reared and housed in cages:

• 20 more eggs per bird per year
• 3.9% higher rate of egg production
• 5.5% less feed required per dozen eggs

- 0.3% fewer large blood spots
- 1.9% fewer cracked eggs
- 3.2% greater livability
- 70 cents more income over feed and chick cost per bird housed.[19]

Such studies and farmers' experience show that there is nothing inefficient or obsolete about farm systems that afford animals more space, exercise, and freedom to move about. They pay off in several ways. Farmers need not pay for costly "health programs," growth stimulants, antibiotics, and other artificial means of propping up productivity depressed by factory conditions. With a much lower capital investment, these farmers have fewer dollars going out to banks, equipment manufacturers, and utility companies. Moreover, the farmers are not locked into specialization, but have flexibility in deciding what to produce.

Instead of policies that promote factory animal production, we need policies that allow the production advantages of traditional methods to come to the fore and enable farmers to make a good living using them. Among such policies should be:

1. Regulations governing livestock production systems and methods. These could benefit animals, farmers, and consumers by requiring that animals be reared with a genuine concern for their health and welfare. Regulatory emphasis on sound husbandry techniques would eliminate the need for drugs, complex management, and isolation in expensive, crowded buildings. The need for greater amounts of specialized labor and the increased space per animal required to conform to the regulations should give small, independent farmers a better position in their competition against large, investor-owned agribusiness factories.

In Europe, where there is stronger public opposition to factory methods on the grounds of poor animal welfare, several countries have regulations aimed at ensuring humane facilities and practices in animal farming. Swedish law, for example, forbids the transport of calves under two weeks of age, muzzling of calves to produce white veal, and permanent tethering of pigs. Swedish law also contains standards governing lighting, ventilation, and space allowances for farm animals, and any persons having animals in their care must allow inspection of their animals and facilities. West German law authorizes its department of agriculture to regulate cages, tethering, and other restrictions on animals' mobility as well as lighting, ventilation, feeding equipment, and other factors. In Switzerland, a new law was established recently by referendum that "forbids detention which is manifestly contrary to the principles of the protection of animals, notably, certain forms of keeping animals in cages and in constant darkness."[20] Under this law, the Swiss cantons can phase out calf stalls and cages for pigs and poultry over the next ten years. Similar laws are being proposed in West Germany, the Netherlands, and Austria. In

Britain, a special committee was appointed by Parliament in 1964 to "examine the conditions in which livestock are kept under systems of intensive husbandry and to advise whether standards ought to be set in the interest of their welfare, and if so what they should be."[21] The "Brambell Committee" recommended, among other things, that debeaking, tail docking, and tethering be prohibited. It also recommended standards concerning cage or stall size, density, lighting, exercise, and other conditions. So far, however, these recommendations have not been given the force of law by the British government.

We would urge some regulation to protect animals immediately from the worst abuses while factory technology persists, but we think basic changes in agriculture and diet offer the only lasting, effective solutions.

2. New priorities in agricultural research and technological development. We should aim in the long run to phase out factory animal production altogether and replace it with more efficient, civilized food production and delivery systems that ensure a healthier, more varied diet. We will go into some of these more fully in the next sections, but we wish to stress here that new priorities should be established at once. We should no longer waste expertise on developing "super" cows and inflatable chickens for agribusiness. USDA and the land grant college complex should be put back on the job of developing systems and methods that are best for consumers, farmers, and the environment.

At the same time, agricultural experts should pay much more attention to farmers' innovations in livestock systems. While many university experts are interested in farmer-designed systems and help in improving them, there is a growing tendency to discount them as crude and unsophisticated. They are not, after all, ordinarily made of expanded metal, extruded aluminum, pumps, motors, switches, and other components put out by experts' favorite agribusiness support companies. They are not, in other words, properly "engineered" for mass production by an agribusiness manufacturer.

Some farmer-built systems use the simplest of materials, require the least labor and energy, and offer a high degree of animal comfort and freedom. Several examples stand out from our survey of farms and magazines, but one of the best was developed by an Illinois farmer who "thinks pigs come first in hog facilities."[22] Instead of restrictive farrowing crates, concrete slabs, and slatted floors, each of his sows and her litter has a "suite" that allows a varied environment and freedom of movement to find comfort. The young pigs can crawl into a wooden "hover" where bedding keeps them dry and, in cold weather, a light bulb provides extra warmth. At nursing time, the pigs join the sow on straw bedding in a dirt-floored farrowing stall. Unlike ordinary stalls, these have only one guard rail eight inches from the floor so the sow can step in and out of the nursing area. "It's not hard to see the first rail on the crate does the work for the sow, the rest just regiments her," says the designer/farmer.[23] After nursing, the sow can take a short walk outside to an outdoor "sunporch" where she has a dunging area. His building has no furnaces or fans, only small gas heaters over each sow that go on in very cold weather. Because his sows dung over outside pits, air in the building is fresh and the labor required to clean twenty-one stalls amounts to about half an hour—done once every other farrowing![24] So much for the "superior efficiency" of high-tech, controlled-environment, total-confinement pig factories.

3. Encourage farming closer to consumers. With food prices rising, quality dropping, and variety narrowing at supermarkets, people are beginning to look elsewhere for food. And much to the chagrin of agribusiness, they are beginning to find it. Roadside and farmers' markets are reviving in and near cities around the country. Charleston, Savannah, New York City, Boston, Seattle, Washington, D.C., Hartford, Chicago, and other cities have successfully restored farmers' markets that enable local farmers to earn more for their labors without raising prices to consumers.[25] Although these markets benefit all economic classes, they are of particular value to the inner-city poor whose neighborhoods often contain few grocery stores and fresh-produce stands. Congress made a token effort to aid this renaissance when it passed the Direct Marketing Act of 1976, but the $1.5 million appropriated is a drop in the bucket compared to the millions in subsidies that go out to agribusiness methods of food production and delivery.

To supply these markets, much more could be done to encourage farming near cities. According to USDA's Economic Research Service, there are still large acreages of cropland reasonably close to most central cities.[26] Near many metropolitan areas, significant agricultural production goes on now, especially vegetable and fruit production.[27] But there is a tendency for commercial development and municipal taxing policies to drive local farmers farther and farther away from markets and consumers. Hawaii, Vermont, Oregon, California, and the Canadian western provinces have established land-use plans designed to protect agricultural land from these forces.[28] Maryland, Massachusetts, New Jersey, and other states have programs that preserve farmland by buying out the "development rights" to the land.[29] Under these plans, the farmer can get money for the development value (the very temptation that forces many to sell out) of their land, yet still stay on it and produce food.

The protection of farmland near cities and the provision of direct markets should enable farmers to make a living producing a broader variety of cash crops instead of being stuck with animal production. But there will still be some demand for animal products while diets and appetites are shifting, and local, independent farmers could supply it without using factory methods. Small-scale poultry and dairy production using more traditional methods could make a comeback with the growth of urban-fringe farming and direct marketing. State agricultural authorities could encourage these farms by eliminating the present agribusiness-serving obstacles imposed by grading and inspection rules. High-quality animal products could be better ensured by thoroughly inspecting and grading farms and production methods rather than by token inspection of the end products at some giant, centralized distribution center.

Revival of food production near population centers could have other benefits as well. Less energy would be spent on shipping, storing, and processing food; less energy would be spent as we shift to less capital-intensive methods. Farmers using more labor-intensive methods would have a large labor pool nearby to draw from; many of our unemployed concentrated in cities could have some relief from poverty, isolation, and boredom by doing farm work—especially at planting and harvesting time.

Understandably, many farmers will look upon these proposals with fear and scorn. Many are intent on producing food exactly as they please and expect USDA and the food corporations to shove it down consumer throats. But consumers are getting more and more conscious of food, diet, and the environment and they are demanding changes. Farmers who really want to stay on their land and make a living raising food will have to become more attentive to these demands.

Farm people have become victims of their own isolation and independence. Many have contempt rather than regard for what goes on in cities where the consumers live, and many are loath to organize or join farmer associations. With these views, it's no wonder that the 3.6 percent of Americans who farm are an endangered species while agribusiness encroaches on their habitat. According to Chuck Frazier of the National Farmers Organization (NFO), "Farmers aren't sufficiently willing to cooperate by joining together and stay on a campaign long enough to get success in pricing their products. They have been burned too many times by organizers and individuals who sought power as self-appointed leaders." He also cites two general developments coming out of this broad discontent:

> They distrust government and many organizations, especially the leadership of government agencies and organizations. This attitude seems to be growing. Secondly, when they do organize, farmers have recently tended to move into specialized commodity groups. As they do so, they dissipate their broad collective strength and stir up competition between different commodity producer groups.[30]

The natural forces of commodity marketing are the primary cause of this sectarianism and specialization in agriculture. With hard times nearly always in sight for independent farmers, they tend to turn to the one crop or type of livestock operation that provides the most security. Then they join with other producers of the same commodity to fight for subsidies, price supports, bans on imports, and other forms of Band-Aid protectionism that they hope will keep their commodity profitable for years to come. The "check-offs" voted in by cattle, hog, and sheep producers are a good example of this kind of tunnel vision. Rice growers, potato producers, and cotton and wheat farmers all have similar commodity promotion campaigns, even though the demand for their products is relatively stable and constant. While farmers foot the bills for these campaigns, the advertising, media, and public relations firms that have the accounts are laughing all the way to the bank, where broad smiles greet them. In the final analysis, neither farmers nor consumers gain much benefit from the millions that change hands. It is ironic that farmers, when they are losing money, are among the loudest howlers at "government controls," yet they dare not whisper about the immense power that commodities traders and other agribusiness interests have over them.

Independent farmers who expect to stay in farming will have to snap out of their anarchism/individualism/isolationism and wake up to consumer, environmental, and other popular concerns. With a truly progressive attitude about food and environmental issues, farmers could gain a fair amount of muscle in coalition with groups working on these issues. Farmers who ignore these trends, or who fight for the narrow goal of making the agricultural status quo more profitable, can expect deepening powerlessness and an increasing trend toward expensive, complicated farming as agribusiness promotes its same old self-serving technology, and government slaps on controls in response to consumer concerns for the environment and food quality.

Toward a Healthier, Happier Future

Agriculture was a great advance for humans, but it spelled doom for much of the rest of the planet. Ten thousand years of human control over the soil has turned good land into deserts, monocultures, and weed patches. To feed ourselves—now over 4 billion strong and bound to double early in the twenty-first century—we destroy ecosystems with bulldozers, plows, and pesticides.[31] Other beings, their habitats destroyed, have neither food nor shelter. Those that survive by adapting to our crops become pests and targets for our poisons, traps, and guns.

Progressive changes in our world view are necessary if we are to maintain a habitable planet. We should strive to limit the extent of our exploitation over animals and soil. While maintaining our hard-won level of civilization, we must extend the scope of moral consideration beyond our own species. We must limit the use of technologies that advance our own short-term success at too great a cost to other creatures and, in the long run, to ourselves.

If these proposals appear far-fetched to some, that may be because the habit of maximizing short-term profit is hard to break. But we can halt the trends that point to ominous ends and start planning an improved quality of life. At the risk of seeming utopian, we propose the following:

1. We should stop expecting agricultural technology and food production to be the solutions to the world's ills. There are those who would intrude even deeper into seas, swamps, jungles, mountains, and deserts to create more cropland to feed people. The problem is not that we don't have enough food to go around. We do. The world food problem exists because of inequities in distribution of food and resources that have grown out of unfair notions about people and property.[32] A few have too much while a great many have too little. Inevitably, those who have too little try to achieve the standard of those who have more than enough. As they do so, social strains occur, followed by increased demands on the environment as human levels of expectation and consumption rise. Instead of pushing for greater agricultural productivity, we should push for ethical, political, economic, and social progress that will equalize distribution of what we take from the earth.

2. Once we have done so, we could reduce human population humanely through family planning. High birthrates are directly related to poverty, malnutrition, and high infant mortality.[33] When a society begins to enjoy a better quality of life and sees its children live to maturity, people feel more secure and hopeful about life, and they desire fewer children as a result. Probably the surest way to head off predicted population increases and their toll on the environment is to divide more equally the world's food, health care, and beneficial technology within the next generation or two.

3. As equitable economic and social institutions are inaugurated and population growth reversed, we could undo much of what we have done over the ages. Many parts of the earth have been overpopulated for centuries. Many areas are so inhospitable to humans that they should be off-limits rather than "conquered" and modified for permanent human habitation. An ethical land-use policy of "not for humans only" could end much of our competition with other species. Much of the earth's land now exploited for food and materials could be freed if humans abandoned animal agriculture. People now engaged in livestock production could be encouraged to turn to other means of livelihood. Over time, millions of acres of land could be returned to wilderness and natural diversity.

4. We should advance new ideologies and technologies of food production. Industrial technology in the hands of profiteers has allowed them to take over agriculture and give us corporate and factory farming. In response to these problems, many reformers favor a return to small, family farms. While self-sufficiency and small size may be beautiful compared to the corporate model, this return to unyielding individualism and simple tools will not provide the answers. The technological solutions

lie in employing the best that *each* scale of food production has to offer. Some people think that mechanization is the root of all evil and that it should be wholly replaced by small-scale, labor-intensive farming methods; others ignore what mechanization does to food quality and blindly advocate labor-eliminating machines as the ultimate liberators of all humankind. Sane food production will require both, in various combinations, depending on the crop and conditions. Factory farming shows that animals don't thrive on mechanical care. Juicy palatable fruits and vegetables don't survive agribusiness's mechanical pickers, so tough, bland ones are developed. But this doesn't mean that we must plant and harvest field crops by hand.

More people could be involved in food production in the future. While only 3.6 percent of the population farm for a living, polls indicate that over half of all Americans grow some of their food and millions of others would like to if they had access to land and assistance.[34] There may be more to this than a fight against rising food prices; there are emotional satisfactions from tilling the soil. People should not be separated from food production and alienated from the environment by either ideology or technology. Future systems of food production should allow people to participate according to their time and skills. Gardening, urban greenhouse farming, urban fringe farms, farmers' markets, and food cooperatives would provide opportunities if people had time, convenient transportation, and economic security to work in them more frequently. For large-scale production of fruits, vegetables, and staples such as cereals and legumes, we ought to consider collective farms. On these farms, a wide range of people could participate in the civic work of food production. Full-time farmer-residents could provide expertise and year-round supervision and could direct long-range planning and operations. Other specialists could be mobile, offering their skills on tours to many farms in different regions. Much of the low-skill work could be done by anyone who has the time and the desire to do outdoor work.

The popularity of urban gardening indicates that many city people enjoy such work as a diversion from office and factory drudgery. Arrangements with employers and other incentives could make it possible for those people to do similar work on the public's farms. These farms could distribute the responsibility and pleasure of food production more evenly throughout society. People could get to know other regions of the country, other kinds of work, and other kinds of people. In the process, urban-rural alienation could be reduced and the quality of rural life enriched by agricultural production that emphasizes sociality, public interest, and the sharing of skills, tools, and labor instead of feed conversion ratios and corporate profits.

Agriculture, our most vital activity, should also be a humane and rewarding occupation.

NOTES

Chapter 1, pages 1–19

1. For poultry industry history, see generally Mack O. North, "Startling Changes Ahead in Production Practices"; Ray A. Goldberg, "Broiler Dynamics—Past and Future"; George E. Coleman, Jr., "One Man's Recollections Over 50 Years"; R. Frank Frazer, "Strategy for the Future"; and "Gordon Johnson Remembers"; all in *Broiler Industry,* July 1976.
2. "How Egg Industry Changed During the Last 20 Years," *Poultry Digest,* May 1978, p. 232.
3. Ibid.; Mack O. North to J. B. Mason, September 18, 1976; and "90% of U.S. Layers Are Housed in Cages," *Poultry Digest,* July 1978, p. 363.
4. Interview with a hatchery manager, Massachusetts, November 7, 1976.
5. North, "Startling Changes," p. 98.
6. J. L. Krider, "Special Problems with Swine in U.S. Midwest Controlled or Semi-controlled Environments (Confinement)," *Proceedings of the III World Conference on Animal Production,* ed. R. L. Reid (Sydney: Sydney University Press, 1975), p. 425.
7. Jackie W. D. Robbins, *Environmental Impact Resulting from Unconfined Animal Production,* Environmental Protection Technology Series (Washington, D.C.: Office of Research and Development, U.S. Environmental Agency, March 1978), p. 9.
8. A. Jensen, et al., *Management and Housing for Confinement Swine Production,* University of Illinois, College of Agriculture, Cooperative Extension Service Circular 1064 (Urbana-Champaign, Illinois: November 1972), p. 3; V. Rhodes and Glen Grimes, "A Study of Large Hog Farms," *Hog Farm Management, 1976 Pork Producers' Planner,* December 1975, p. 15; and "Disease Major Limiting Factor," *Hog Farm Management,* May 1976, p. 30.
9. *Farm Journal,* December 1975, p. 33; and "Producers Predict More Confinement for Corn Belt Hogs," *National Hog Farmer,* August 1975, p. 5.
10. Advertisement for Farmstead Industries, Waterloo, Iowa, in *Hog Farm Management,* April 1975, p. 1.
11. Ibid.
12. Robbins, *Environmental Impact,* p. 9.
13. Precise figures on the number of "milk-fed" veal calves produced each year

147

are not available because neither state nor federal authorities require reports. Our estimate is based on an interview with a veal producer, Connecticut, September 1974, and an article: "Veal Calves and Factory Farming," *Report to Humanitarians* (St. Petersburg, Fla.: Humane Information Services), no. 43, March 1978, p. 1.

14. Interview with a veal producer, Connecticut, September 1974.
15. Ibid.
16. *Agricultural Statistics 1978* (Washington, D.C.: U.S. Government Printing Office, 1978), table 454, p. 306. Cattle on grain or other concentrates on January 1, 1973, numbered 14½ million.
17. "Feedlot Numbers Shrink," *Beef,* April 1977, p. 28.
18. Ibid.
19. John Dawson, "Sheep Come Back," *Confinement,* March 1978, p. 9.

Chapter 2, pages 21–33

1. M. Kiley, "The Behavioural Problems that Interfere with Production in Animals under Intensive Husbandry," *Proceedings of the III World Conference on Animal Production,* ed. R. L. Reid (Sydney: Sydney University Press, 1975), p. 431.
2. Stanley E. Curtis, "What is 'Environmental Stress'?" *Confinement,* June 1976, p. 15.
3. "Tail-biting Is Really 'Anti-comfort Syndrome,'" *Hog Farm Management,* March 1976, p. 94.
4. Neal Black, "Production Drops If Sows Confined," *National Hog Farmer,* Swine Information Service Bulletin no. E-13.
5. Kiley, "Behaviour Problems," p. 431.
6. It is estimated that 80 percent of swine herds in the United States carry atrophic rhinitis and as many as half of the pigs in a herd are infected. The incidence of swine dysentery has increased 88 percent in the past four years because of factory farming's trend toward more extensive movement of feeder pigs. Transmissible gastroenteritis (TGE) is on the rise in continuous farrowing operations. "A Status Report on the Six Most Deadly Hog Diseases," *Successful Farming,* October 1976, pp. H-12, 13. Operators of large pig farms were surveyed recently and asked to list the advantages and disadvantages of their systems; the most often cited (23 percent of respondents) disadvantage was "disease problems." The farms surveyed used a variety of buildings and systems, but 57.2 percent used some type of confinement facilities. "Disease Major Limiting Factor," *Hog Farm Management,* May 1976, p. 30. See also G. W. Meyerholz, "Disease Management in Large Herds," *Confinement,* June 1976, p. 11.
7. "Keep Those Dairy Calves Alive!" *Successful Farming,* April 1977, p. D-2.
8. D. Halvorson, "What You Can Do About Those New and Re-emerging Old Diseases," *Poultry Digest,* February 1975, p. 63.

9. Ralph Vinson, "Using Health to Increase Profits," *Hog Farm Management,* January 1978, p. 72.
10. Stanley N. Gaunt and Roger M. Harrington, eds., *Raising Veal Calves* (n.p.: The Massachusetts Cooperative Extension Service, n.d.), no. 106, p. 4.
11. Interview with a veal producer, Connecticut, September 1973.
12. Eldon W. Kienholz, "Vitamin Deficiencies in Chickens and Turkeys," *Poultry Digest,* April 1976, p. 142.
13. Mack O. North, "Catching Up on Smaller Profit Leaks," *Broiler Industry,* June 1976, p. 41.
14. John B. Herrick, "Liver Abscesses Are Costly Problem in Most Feedlots," *Beef,* September 1976, p. 29.
15. D. A. Hartman, "What About Vitamin A, D and E Injections?" *Hoard's Dairyman,* April 10, 1976, pp. 500–501.
16. Robert D. Fritschen, "Floors and Their Effects on Feet and Leg Problems in Swine," *Confinement,* June 1976, p. 12; and Gene Johnston, "Boars: Keeping Them on Their Feet," *Successful Farming,* October 1977, p. H-20.
17. Ibid., and Robert D. Fritschen, "Stomp Out Foot and Leg Problems," *Hog Farm Management,* March 1978, p. 78.
18. Richard Wall, "Cage Layer Fatigue," *Poultry Digest,* January 1976, p. 23.
19. Milton Y. Dendy, "Broiler 'Flip-over' Syndrome Still a Mystery," *Poultry Digest,* September 1976, p. 380.
20. Interview with a broiler producer, North Carolina, December 1978.
21. C. I. Angstrom, "Mechanical Failures Plague Cage-layers," *Onondaga County Farm News* (Syracuse, N.Y.), December 1970, p. 13; and Charles W. Howe, "Which . . . Started Pullets or Stunted Pullets?" *Poultry Digest,* July 1976, p. 295.
22. Vernon C. Thompson, "Heat Wave Killing MD, VA Chickens," *The Washington Post,* July 17, 1977.

Chapter 3, pages 35–51

1. Stephen Singular, "Brave New Chickens," *New Times,* April 29, 1977, p. 42; A. Van Tienhoven and C. E. Ostrander, "Short Total Photoperiods and Egg Production of White Leghorns," *Poultry Science,* vol. 55, pp. 1361–64, 1976.
2. "Highlights of Poultry Science Papers," *Poultry Digest,* October 1976, p. 422.
3. Paul Siegel and Bernie Gross, "We're Learning How to Let Bird Defend Itself," *Broiler Industry,* August 1977, p. 42.
4. *Progress in Feeding and Animal Health* (Webster City, Iowa: Woodard Animal Health Premix Division, n.d.), p. 12.
5. "Hog Management: Is No Tail Biting a Success?" *Successful Farming, 1976 Machinery Management Issue,* February 1976, p. A-1.

6. F. D. Thornberry, W. O. Crawley, and W. F. Krueger, "Debeaking Laying Stock to Control Cannibalism," *Poultry Digest,* May 1975, pp. 205–7.
7. Ibid.; and "Best Age for Debeaking is 6 Days," *Poultry Digest,* July 1976, p. 283.
8. Rocky J. Terry, "How to Use Antibiotics Effectively," *Poultry Digest,* November 1975, p. 440.
9. Jack Denton Scott, "The Incredible Edible Egg," *Reader's Digest,* July 1978, p. 55; *Columbia Daily Tribune* (Columbia, Mo.), December 1, 1977; and H. V. Biellier to J. B. Mason, February 8, 1979.
10. H. G. Purchase and J. W. Deaton, "Food from Poultry for Future," *Broiler Industry,* July 1976, p. 360.
11. "Has the Egg Changed?" *Poultry Digest,* November 1977, p. 522.
12. ABC News Closeup, "Food: Green Grow the Profits," Friday, December 21, 1973.
13. "Naked Chick Gets Serious Attention," *Broiler Industry,* January 1979, p. 98.
14. ABC News Closeup.
15. "Tests Show Stress-prone Gilts Are Late Breeders," *National Hog Farmer,* October 1976, p. 98.
16. Thayne Cozart, "AI Cuts Costs, Forces Better Management," *National Hog Farmer,* January 1976, p. 86.
17. "Scientist Studies 'Test Tube Pig,'" *Hog Farm Management,* April 1975, p. 61.
18. *Economic Effects of a Prohibition on the Use of Selected Animal Drugs,* U.S. Department of Agriculture, Economics, Statistics, and Cooperative Service, Agricultural Economic Report no. 414 (Washington, D.C., November 1978), p. 4.
19. *Food Safety and Quality, Use of Antibiotics in Animal Feed:* Hearings before the Subcommittee on Agricultural Research and General Legislation of the Committee on Agriculture, Nutrition and Forestry, U.S. Senate, 95th Congress, 1st Session, September 21–22, 1977 (statement of Dr. Richard P. Novick), p. 108.
20. "What Tells Cattle to Stop Eating?" *Beef,* November 1976, p. 33.
21. Peter Watson, "Boss Cows and Sleepy, Sexy Sheep," *Psychology Today,* December 1975, pp. 93–94.
22. James L. Kerwin, "Well, Son, Cows Eat Old Boxes, Give Milk," *The Detroit News,* April 19, 1976.
23. John Byrnes, "My Dry Sows Go 90 Days Without Feed," *Hog Farm Management,* April 1978, p. 36.
24. Ibid.
25. "Oxidation Ditch Liquid Boosts Efficiency 10%," *National Hog Farmer,* August 1975, p. 30.
26. Mack O. North, "Startling Changes Ahead in Production Practices," *Broiler Industry,* July 1976, pp. 81, 85.
27. Ibid.
28. Mack O. North, "Some Tips on Floor Space and Profits," *Broiler Industry,* December 1975, p. 24.

29. Veryl Sanderson and Bill Eftink, "Their Dream House Nearly Drove Them to the Poor House," *Successful Farming,* August 1978, pp. B-6–7.
30. Ibid.
31. C. E. Ostrander and R. J. Young, "Effects of Density on Caged Layers," *New York Food and Life Sciences,* July–September 1970, pp. 5–6.
32. R. L. Kohls, "Sounding the Alarm for Ag Research," *Confinement,* June 1977, p. 4.

Chapter 4, pages 53–69

1. William J. Stadelman, "Lower Quality with Yolk Color Increase," *Egg Industry,* December 1975, p. 24.
2. Roger J. Williams et al., "The 'Trophic' Value of Foods," *Proceedings of the National Academy of Science, U.S.A.,* 70:3 (March 1973), 710–13; A. Tolan, et al., "The Chemical Composition of Eggs Produced Under Battery, Deep Litter and Free Range Conditions," *British Journal of Nutrition,* 30: 181 (March 1974), 185.
3. M. A. Crawford, "A Re-evaluation of the Nutrient Role of Animal Products," *Proceedings of the III World Conference on Animal Production,* ed. R. L. Reid (Sydney: Sydney University Press, 1975), p. 24.
4. Ibid., p. 26.
5. William J. Stadelman, "Old-time Flavor: New Injectables Possible," *Broiler Industry,* April 1975, p. 79.
6. Gregory Leonardos, "Brand Life May Depend on Unique Flavors," *Broiler Industry,* October 1976, p. 33.
7. Ibid.
8. Joe Dan Boyd, "Coming: A New Wave of Hard-hitting Insecticides," *Farm Journal,* October 1976, p. 20.
9. *Drugs in Livestock Feed,* vol. I: *Technical Report* (Washington, D.C.: Office of Technology Assessment, June 1979), p. 3.
10. "Industry-Government 'Self-help' Sulfa Campaign Underway," *Northeast Regional Information Office Newsletter,* Food Safety and Quality Service, U.S. Department of Agriculture (New York: June 15, 1978), p. 1.
11. "Teat Dips Due for FDA Scrutiny," *Dairy Herd Management,* April 1976, p. 28.
12. "New Treatment Boosts Pigs Per Litter," *Farm Journal,* March 1976, p. Hog-2.
13. *Drugs in Livestock Feed,* vol. I, p. 3.
14. Advertisement for Esmopal in *Poultry Digest,* October 1977, p. 497.
15. "Drug Use Guide: Swine," Bureau of Veterinary Medicine, Department of Health, Education and Welfare, FDA 76-6012 (Rockville, Md.: B.V.M., Industry Relations Branch, May 1976), p. 1.
16. Monroe Babcock, "Shrinking Egg Market Is Our Own Fault," *Egg Industry,* January 1976, pp. 29–30.

17. "Fresh Chicken: It's a Long Way from Farmer Jones to Frank Perdue," *Consumer Reports,* May 1978, p. 255.
18. Earl Ainsworth, "Now a U.S. Ham That Will Cut into Competition," *Farm Journal,* September 1976, p. Hog-8.
19. "PBB: Michigan Contamination Continues," *Guardian,* May 4, 1977, p. 2.
20. Daniel Spitzer, "Is the Poisoning of Michigan Just the Start?" *Mother Jones,* May 1977, p. 18.
21. *Problems in Preventing the Marketing of Raw Meat and Poultry Containing Potentially Harmful Residues* (Washington, D.C.: Comptroller General of the United States, April 17, 1979), p. i.
22. P. F. McGargle, "The Slaughterhouse Scandal," *World Magazine,* June 25, 1977, p. M-5.
23. "Feeder Pig Spokesman Asks Shipping, Starting Research," *National Hog Farmer,* May 1976, pp. 10–11.
24. Ralph D. Wennblom, "Government Holds Firm on Sulfa Drugs," *Farm Journal,* mid-February 1978, p. Hog-13.
25. John Byrnes, "Let Them Eat Waste," *Hog Farm Management,* June 1978, p. 50.
26. "Little Effect of British Antibiotic Restrictions," *National Hog Farmer,* March 1977, p. 142.
27. Rex Wilmore, "Many Problems in British Antibiotic Policy," *National Hog Farmer,* October 1975, p. 18.
28. "Washington Report: Illegal Drug Sales," *Successful Farming,* February 1976, p. 4.
29. "FDA Crackdown on Illegal Use of Dysentery Drug," *National Hog Farmer,* September 1976, p. 5.
30. "Inspection—A Refreshing Challenge!" *Broiler Industry,* November 1975, p. 50.
31. Carl Cason, "Automation Esthetics and Productivity," *Broiler Industry,* December 1975, pp. 21–22.
32. *Problems in Preventing the Marketing of . . . Meat . . . Containing . . . Harmful Residues,* pp. 7–13.
33. Daniel Hays, "Meat Packers Admit Bribes," *The Daily News* (New York), May 24, 1977; Peter Schuck, "The Curious Case of the Indicted Meat Inspectors," *Harper's,* September 1972, p. 81. The problems of industry influence and corruption of meat inspectors have been investigated and reported by a team of workers under Ralph Nader. Their report, written by Harrison Wellford, was published as *Sowing the Wind* (New York: Grossman Publishers, 1972).
34. "Angelotti Resigns as Head of Food Inspection Service," *Poultry Digest,* September 1978, p. 488; "Embattled USDA Chief Resigns," *National Hog Farmer,* September 1978, p. 106.
35. *The Animal Health Institute Annual Report 1975* (Washington, D.C.: Animal Health Institute, 1975), p. 17.
36. Neal Black, "Prescription for High Costs, Less Drug Help," *National Hog Farmer,* September 1976, p. 36.
37. Gary L. Cromwell, "Antibiotic Feed Additive Benefits Documented," *National Hog Farmer,* April 1978, pp. 42, 46; and "Antibiotics," *NCA*

Digest (National Cattlemen's Association) I:9 (August 1978), 3.

38. John Russel, "Antibiotics in Feed: The First Punch Has Been Thrown," *Farm Journal,* June/July 1977, p. Hog-20; and Roland C. Hartman, "Will Antibiotics Use Survive?" *Poultry Digest,* October 1976, p. 395.

39. R. F. Wilson, "Hoginformation: Cutting Costs," *National Hog Farmer,* August 1976, p. 89.

40. *Economic Effects of a Prohibition on the Use of Selected Animal Drugs,* U.S. Department of Agriculture, Economics, Statistics, and Cooperative Service, Agricultural Economic Report no. 414 (Washington, D.C., November 1978), p. ii; and John McClung, "Washington Report: USDA Reported Drug Ban Beneficial," *Hog Farm Management,* December 1978, p. 6.

41. Ibid.

Chapter 5, pages 71–81

1. Frances Moore Lappé, *Diet for a Small Planet,* rev. ed. (New York: Ballantine Books, 1975), p. 13.

2. Ibid., p. 17.

3. "Eleven Thousand Dead Pigs on 26 Farms," *Successful Farming,* January 1977, p. A-2.

4. "Keep Those Dairy Calves Alive," *Successful Farming,* April 1977, p. D-2.

5. Mack O. North to J. B. Mason, March 20, 1979. Mortality is estimated at .75 percent and culling at .25 percent per month, for a total monthly loss of 1 percent of the flock. Assuming an eighteen-month laying cycle with one force-molt causing loss of 10 percent of the flock, total losses for the entire cycle will be 28 percent. Data from California egg operations, both those using force-molting and those not, indicate an average loss from mortality and culling of 1.25 percent per month.

6. Tony J. Cunha, "Productivity Must be Increased," *Hog Farm Management,* February 1979, p. 34.

7. D. G. Fox and F. R. Black, "New Tool Pulls It All Together: Systems Analysis," *Confinement,* May 1976, p. 12.

8. "Coccidiosis Most Costly of Poultry Diseases," *Poultry Digest,* November 1976, p. 442.

9. "Pig Health Losses Total $187 Million," *Farm Journal,* September 1978, p. Hog-2.

10. Bill Miller, "Brucellosis," *Successful Farming,* August 1978, p. B-18.

11. *Animal Health Institute Fact Book* (Washington, D.C.: Animal Health Institute, n.d.), p. 2.

12. Ibid.

13. *A Research Program to Reduce Losses Due to Transportation of Livestock,* Executive Communication no. 2058 (Washington, D.C.: Department of Agriculture, Agricultural Research Service, May 1977), p. 1.

14. John R. Dawson, "Death Enroute to Market," *Confinement,* June 1977, p. 14.
15. *Agricultural Statistics, 1978* (Washington, D.C.: U.S. Government Printing Office, 1978), table 582, p. 405, calculated by dividing pounds condemned (live weight) by 4.377 pounds—the average live weight of chickens, ducks, and turkeys; mammals from *Statistical Summary: Federal Meat and Poultry Inspection for Fiscal Year 1976* (Washington D.C.: Department of Agriculture, Animal and Plant Health Inspection Service, January 1977), table 1, p. 1.
16. *Agricultural Statistics, 1978,* table 526, p. 357; *Statistical Summary,* table 5, p. 5.
17. Ibid., table 582, p. 405 (average for years 1975–1977).
18. John B. Herrick, "Liver Abscesses Are Costly Problem in Most Feedlots," *Beef,* September 1976, p. 16.
19. Harrison Wellford, *Sowing the Wind* (New York: Grossman Publishers, 1972), p. 82.
20. J. T. Reid, "Comparative Efficiency of Animals in the Conversion of Feedstuffs to Human Foods," *Confinement,* April 1976, p. 23.
21. "Oxidation Ditches Save Money," *Farm Journal,* October 1976, p. Hog-24.
22. W. L. Roller, H. M. Keener, and R. D. Kline, "Energy Costs of Intensive Livestock Production" (St. Joseph, Mich.: American Society of Agricultural Engineers, June 1975), paper no. 75-4042, table 7, p. 14.
23. Ibid., table 9, p. 15.
24. Ibid., tables 7 and 8, pp. 14, 15.
25. Ibid., p. 6.

Chapter 6, pages 83–95

1. *Food Safety and Quality, Use of Antibiotics in Animal Feed:* Hearings before the Subcommittee on Agricultural Research and General Legislation of the Committee on Agriculture, Nutrition and Forestry, U.S. Senate, 95th Congress, 1st Session, September 21–22, 1977 (statement of Dr. Richard P. Novick), pp. 17–29.
2. Ibid.
3. Raymond C. Loehr, *Pollution Implications of Animal Wastes—A Forward Oriented Review,* Water Pollution Control Research Series (Washington, D.C.: Office of Research and Monitoring, U.S. Environmental Protection Agency, 1968), p. 26 and table 7, p. 27. A hen in confinement produces about .39 pound of manure per day; 60,000 hens produce 163,800 pounds of manure per week, or 81.9 tons. One pig produces about 17.36 pounds of mixed manure and urine each day, or 121.52 pounds each week.
4. H. A. Jasiorowski, "Intensive Systems of Animal Production," *Proceedings of the III World Conference on Animal Production,* ed. R. L. Reid

(Sydney: Sydney University Press, 1975), p. 384; Jackie W. D. Robbins, *Environmental Impact Resulting from Unconfined Animal Production,* Environmental Protection Technology Series (Cincinnati: U.S. Environmental Protection Agency, Office of Research and Development, Environmental Research Information Center, February 1978), p. 9.

5. Loehr, *Pollution Implications,* pp. 22–23.
6. Ibid., p. 48.
7. Ibid., pp. 48–49.
8. "Products for Odor Control Fail Tests," *National Hog Farmer,* February 1979, p. 24.
9. Letter in "Hoginformation Please," *National Hog Farmer,* December 1975, p. 36.
10. *Lawrence County Record* (Mt. Vernon, Mo.), September 28, 1978; Ralph Watkins, "Pollution Issue Used to Block Large Unit," *National Hog Farmer,* July 1978; and Ralph Watkins, "Odor Complaints Force Custom Feedlot Shutdown," *National Hog Farmer,* April 1978.
11. "Handling Waste Disposal Problems," *Hog Farm Management,* April 1978, pp. 16, 18.
12. Ibid., p. 17.
13. *Epidemiological Aspects of Some of the Zoonoses,* DHEW Publication no. CDC 75-8182 (Atlanta: U.S. Department of Health, Education and Welfare, Center for Disease Control, Office of Veterinary Public Health Services, November 1973); W. T. Hubbert, W. F. McCulloch, and P. R. Schnurrenberger, *Diseases Transmitted from Animals to Man* (Springfield, Ill.: Charles C. Thomas, 1975).
14. Vivian Wiser, *Protecting American Agriculture: Inspection and Quarantine of Imported Plants and Animals,* Agricultural Economic Report no. 266 (Washington, D.C.: U.S. Department of Agriculture, Economic Research Service, July 1974); and "Protecting America's Animal Health," flyer (Washington, D.C.: U.S. Department of Agriculture, Veterinary Services, Animal and Plant Health Inspection Services, July 1975), p. 2.
15. Bill Miller, "Brucellosis," *Successful Farming,* August 1978, p. B-18.
16. John A. Rohlf, "We Have the Tools to Eradicate Brucellosis," *Farm Journal,* March 1976, p. Beef-14; "Showdown on Brucellosis," *Farm Journal,* January 1976, p. Beef-9.
17. Neal Black, "145-year Cholera Battle Ends," *National Hog Farmer,* February 1978, p. 50.
18. Gilberto S. Trevino, "Foreign Animal Disease Control Programs in the United States," *Journal of the American Veterinary Medical Association,* September 15, 1975, p. 459.
19. "Pseudorabies Eradication Plan Drafted," *National Hog Farmer,* March 1977, p. 136.
20. John Byrnes, "Demand Grows for PRV Vaccine," *Hog Farm Management,* May 1977, pp. 18, 20.
21. "Area Depopulation Plan Suggested for Dominican," *National Hog Farmer,* December 1978, p. 34.
22. *ASCS Commodity Fact Sheet,* 1978–79 Dairy Program (Washington, D.C.:

U.S. Department of Agriculture, Agricultural Stabilization and Conservation Service, January 1979).

23. Ibid.

24. Michael McMenamin and Walter McNamara, "Milking the Public," *Inquiry,* November 13, 1978.

25. Ibid.

26. Joe Belden and Gregg Forte, *Toward a National Food Policy* (Washington, D.C.: Exploratory Project for Economic Alternatives, 1976), p. 81. For further information on the dairy lobby see: Michael T. McMenamin and Walter McNamara, *Milking the Public: Political Scandals of the Dairy Lobby from LBJ to Jimmy,* tentative title (Chicago: Nelson-Hall, to be published in 1980); for further information on the federal milk price support system, see Paul W. MacAvoy, ed., *Federal Milk Marketing Orders and Price Supports* (Washington, D.C.: American Enterprise Institute for Public Policy Research, 1977).

27. Karen Brown, "Investment Credit Battle Won," *Hog Farm Management,* December 1978, p. 14.

28. "SBA and FmHA Loan Programs Violate 1977 Farm Bill," Center for Rural Affairs Newsletter (Walthill, Neb.), August 1979.

29. "Poultry Scientists Receive Awards," *Broiler Industry,* September 1976, p. 68.

30. H. G. Purchase and J. W. Deaton, "Food from Poultry for Future," *Broiler Industry,* July 1976, pp. 36C, 43.

Chapter 7, pages 97–111

1. Radoje Nikolitch, *Family-size Farms in U.S. Agriculture* (Washington, D.C.: U.S. Department of Agriculture, Economic Research Service, February 1972), p. iii.

2. Peter M. Emerson et al., *Public Policy and the Changing Structure of American Agriculture* (Washington, D.C.: U.S. Government Printing Office, August 1975), pp. 23–24.

3. Eugene Ross, "Our Dilemma," *Confinement,* October 1977, p. 24.

4. Emerson et al., *Public Policy,* p. xii.

5. David W. Wilson, in "Hear Me . . . Why Farmers Get Big," *Farm Journal,* September 1976, p. 14.

6. Emerson, *Public Policy,* p. 4.

7. "Making Farmers Disappear," *The Progressive,* November 1976, p. 9; and Emerson, *Public Policy,* p. 8.

8. Ibid.

9. "Peripheral Birds More Numerous Than Realized," *Poultry Digest,* November 1976, p. 446.

10. Ibid.

11. Don Paarlberg, "Land Prices Are Running a Fever," *Farm Journal,* April 1977, p. 17.

12. Neal Black, "Let's Give USDA to Do-gooders, Gardeners," *National Hog Farmer,* August 1976, p. 26.
13. "New 'Food Policy' Report Overlooks Past Experiences," *Beef,* April 1977, p. 12.
14. Debra Switzky, "Seedstock, Confinement Trends Forecast," *National Hog Farmer,* February 1977, pp. 103–4.
15. Ibid., "Industry Voices," *Hog Farm Management,* June 1978, p. 59.
16. From "The Family Farm as a Multinational Business," by Stephen Singular. Copyright © 1975 by News Group Publications, Inc. Reprinted with the permission of *New York* magazine.
17. "Egg Business Roundup: The Big Boys Are Doing the Expansion," *Poultry Tribune,* February 1975, p. 8.
18. Ibid.
19. "Four Largest Do 18% of 1975 Volume," *Broiler Industry,* March 1976, p. 16.
20. Forest V. Muir to J. B. Mason, February 20, 1975; Kenneth E. Wing and Frank D. Reed, *Costs and Returns on Maine Broiler Farms,* Bulletin 662, Maine Agricultural Experiment Station (Orono: University of Maine, September 1968); Edward S. Micka, *The Competitive Position of the Maine Poultry Industry,* Bulletin 732, Life Sciences, and Agricultural Experiment Station (Orono: University of Maine, November 1976).
21. C. L. Quarles and D. D. Caveny, "How You'll Do Business by 1985!" *Broiler Industry,* July 1976, p. 34.
22. Ralph L. Baker, "One Man's View: Where We'll Be in the 1980's," *Egg Industry,* December 1975, pp. 19–20.
23. Edward H. Covell, "Enough Is Enough," *Broiler Industry,* June 1976, p. 54.
24. "Automation Can Halve Plant Labor in 3 Years," *Broiler Industry,* January 1977, p. 72.
25. Eldon Kreisel, Rocheport, Mo., hog producer, quoted in Glenn Grimes, "Role of Large Units Is Uncertain," *Hog Farm Management,* August 1977, p. 56.
26. Bob Tuten, "Hogs, Territory Expansion in Pennfield's Growth Plan," *Broiler Industry,* December 1977, pp. 22, 32.
27. "Some Hogmen Quit—Others Keep Expanding," *Successful Farming,* 1976 Planning Issue, December 1975, p. H-12.
28. Ibid.
29. V. J. Rhodes and G. A. Grimes, "The Structure of Pork Production," *Confinement,* September 1976, p. 23.
30. "$27 Million in Escrow to Kill Anti-trust Suit," *Broiler Industry,* September 1977, p. 24; and "Broiler Marketing Group Votes to Disband," *Poultry Digest,* September 1978, p. 488.
31. "EGGMAR Has 'Muscle' as It Starts Central Selling," *Egg Industry,* August 1976, p. 14; and "EGGMAR in Perspective," *Egg Industry,* October 1976, p. 38.
32. "USDA Charges Iowa Beef with Monopoly on Prices," *Beef,* February 1978, p. 79.
33. Joe Belden and Gregg Forte, *Toward a National Food Policy* (Washington,

D.C.: Exploratory Project for Economic Alternatives, 1976), p. 80; Linda Kravitz, *Who's Minding the Co-op?* (Washington, D.C.: Agribusiness Accountability Project, March 1974).

34. Ibid.
35. Bob Doerschuk, "America's Last Monopoly: Milk Monoliths Examined," *Nutrition Action,* January 1976, p. 8.
36. Linda Kravitz, *Who's Minding the Co-op?* (Washington, D.C.: Agribusiness Accountability Project, March 1974).
37. "Beeferendum Effort," *Confinement,* October 1978, p. 9.
38. "Beef Referendum Is an Opportunity for Whole Beef Industry," *Beef,* July 1977, p. 6.
39. "Egg Board Budgets $6.3 Million for 1979," *Poultry Digest,* September 1978, p. 488.
40. "UDIA Working Hard to Promote Dairy Products," *Successful Farming,* March 1976, p. A-2.
41. "Swift to Push Chicken Brand," *Broiler Industry,* September 1976, p. 58.
42. Stanley E. Curtis, "Getting the Story Told," *Confinement,* May 1978, p. 18.
43. J. M. Lewis et al., "More on Sheep Flooring," *Confinement,* July-August 1977, p. 6.
44. John Rohlf, "Your Beef Business," *Farm Journal,* December 1978, p. Beef-20.
45. Roland C. Hartman, "Countering with Facts," *Poultry Digest,* March 1978, p. 114.
46. Bayard Webster, "6 Scientists Quit Panel in Dispute Over Livestock Drugs," *The New York Times,* January 23, 1979, p. C-2; "Scientists Quit Anti-biotics Panel at CAST," *Science,* 203 (February 23, 1979), 732.

Chapter 8, pages 113–27

1. *Agricultural Statistics, 1978* (Washington, D.C.: U.S. Government Printing Office, 1978), table 772, p. 556.
2. Boyce Rensberger, "For Most, Beef Is the Staple," *The New York Times,* May 24, 1978, p. C-1.
3. Joe Belden and Gregg Forte, *Toward a National Food Policy* (Washington, D.C.: Exploratory Project for Economic Alternatives, 1976), chap. 4.
4. Nathaniel Altman, "Revising the 'Basic Four,'" *Vegetarian Times,* September/October 1977, p. 9.
5. F. J. Schling and M. C. Phillips, *Meat Three Times a Day* (New York: Richard R. Smith, 1946), p. 54.
6. Frances Moore Lappé and Joseph Collins, *Food First: Beyond the Myth of Scarcity* (Boston: Houghton Mifflin Co., 1977), p. 244.
7. Frances Moore Lappé, *Diet for a Small Planet,* rev. ed. (New York: Ballantine Books, 1975), p. xvii.
8. Ibid., pp. 12–13.

9. *Beef,* January 1979, p. A-3.
10. *The New York Times,* November 3, 1974 (shooting calves); Norma Jane Skjold, "We Need a Red Cross for Starving Cattle," *Farm Journal,* December 1976, pp. 10–12 (starving cattle); "Milk Price Improved," *Hoard's Dairyman,* April 10, 1976, p. 456 (dumping milk).
11. "Climbing the Protein Ladder," *Farm Journal,* December 1978, p. 52.
12. Lappé and Collins, *Food First,* p. 330.
13. Harrison Wellford, *Sowing the Wind* (New York: Grossman Publishers, 1972), p. 103.
14. Ibid.
15. Warren Kester, "How Much Integration for the Pork Industry?" *Farm Journal,* October 1975, p. H-6.
16. Peter M. Emerson et al., *Public Policy and the Changing Structure of American Agriculture* (Washington, D.C.: U.S. Government Printing Office, August 1975), p. xi.
17. Gene Logsdon, "Leave Us Alone and We'll Produce the Food," *Farm Journal,* June/July 1976, p. 16.
18. *Farm Journal,* August 1976, p. 12.
19. *An Enquiry into the Effects of Modern Livestock Production on the Total Environment* (London: The Farm and Food Society, 1972), p. 12.
20. A. J. Koltveit, D.V.M. of Elgin, Ill., in *Confinement,* November/December 1976, p. 3.
21. Page Smith and Charles Daniel, *The Chicken Book* (Boston and Toronto: Little, Brown and Company, 1975), p. 303.

Chapter 9, pages 129–45

1. Frances Moore Lappé and Joseph Collins, *Food First: Beyond the Myth of Scarcity* (Boston: Houghton Mifflin Co., 1977), p. 403.
2. Numerous articles have appeared in technical and lay periodicals, newspapers, and magazines in recent years on the health risks associated with excessive meat and protein consumption. See: Jane E. Brody, "Studies Suggest a Harmful Shift in Today's Menu," *The New York Times,* May 15, 1979, p. C-1; M. A. Crawford, "A Re-evaluation of the Nutrient Role of Animal Products," *Proceedings of the III World Conference on Animal Production,* ed. R. L. Reid (Sydney: Sydney University Press, 1975), p. 21; Patricia Hausman, "Protein: Enough Is Enough," *Nutrition Action,* October 1977; Robin Hur, "Vegetarians for a World of Plenty," *Moneysworth,* March 1979; Phyllis Machta and Michael Jacobsen, "Protein: The 'Buy' Word," *Nutrition Action,* January 1976, p. 4; Hana Marano, "The Problem with Protein," *New York,* March 5, 1979, p. 49. For references to technical publications, see: *Diet Related to Killer Diseases,* III, Hearings Before the Select Committee on Nutrition and Human Needs, U.S. Senate, 95th Congress, 1st Session, March 24, 1977

(statement of Alex Hershaft, Ph.D.), p. 322.

3. Frances Moore Lappé, *Diet for a Small Planet* (New York: Ballantine Books, 1975), p. 41; Patricia Hausman, "Protein: Enough Is Enough."

4. Ibid.

5. *Dietary Goals for the United States,* Committee Print, Select Committee on Nutrition and Human Needs, U.S. Senate, 95th Congress, 1st Session, December 1977.

6. "Vegetarianism Becoming More Popular, Menus Shifting to Match the Growing Demand," *Meat Board Reports* (Chicago: National Livestock and Meat Board) 10:6 (March 28, 1977), 4.

7. Private communication from the Roper Organization, December 1978; "A New Vegetarian Poll," *Vegetarian Times,* January/February 1979, p. 7.

8. George C. Inglett, *Fabricated Foods* (Westport, Conn.: AVI Publishing Co., 1975), p. 4.

9. "How to Match Capital and Labor in the Hog Business," *Successful Farming,* January 1977, p. 9.

10. "Royal Show Demo," *AG Newsletter* (Petersfield, England: Compassion in World Farming), no. 2 (August 1976), p. 2.

11. Joe Belden and Gregg Forte, *Toward a National Food Policy* (Washington, D.C.: Exploratory Project for Economic Alternatives, 1976), p. 48.

12. The American appetite for beef has caused enormous damage to our environment, especially in the West, where millions of acres have been overrun and overgrazed since the late nineteenth century. See Phillip L. Fradkin, "The Eating of the West," *Audubon,* January 1979; James Nathan Miller, "The Nibbling Away of West," *Reader's Digest,* December 1972, p. 107; Sixth Annual Report, Council on Environmental Quality (Washington, D.C., December 1978), p. 212.

13. John Dawson, "No Farrowing Crates, No Antibiotics, No Furnaces, No Fans . . . But the Pigs Come First," *Confinement,* June 1976, p. 7.

14. "Planned Exposure Breaks SMEDI's Hold," *Successful Farming,* March 1976, p. H-16 (Nebraska); Vance Ehmke, "Farrowing Huts a Practical Way to Boost Volume," *Farm Journal,* December 1976, p. Hog-11 (Kansas); John Byrnes, "Slatted Floors Are Starting to Mix with Bedding," *Hog Farm Management,* November 1978, p. 20 (Ohio); and Ron Brunoehler, "I'll Never Have Crates Again," *Successful Farming,* January 1979, p. H-2 (Wisconsin).

15. Dale McKee, "Ten Years of Confinement Raises More Questions Than Answers," *Hog Farm Management,* March 1979, p. 124.

16. Ibid.

17. Ibid.

18. "Outside Pigs More Efficient," *National Hog Farmer,* January 1977, p. 86.

19. Milton Y. Dendy, "Highlights of Poultry Science Meeting Reports: 20 More Eggs for Floor System," *Poultry Digest,* September 1976, p. 366. G. A. Martin et al., "Layer Performance in Cage vs. Non-cage Housing," paper presented by G. W. Morton, Jr., at 65th annual meeting of Poultry Science Association at Kansas State University, Manhattan, Kan., August 2–6, 1976 (abstracted in *Poultry Science* 55: 5 [1976], 2060).

20. Swiss Tiershutz, 1978, section 2, 4th article.
21. *Report of the Technical Committee to Enquire into the Welfare of Animals Kept Under Intensive Livestock Husbandry Systems,* Command Paper 2836 (London: Her Majesty's Stationery Office, 1965), p. A-3.
22. Karen Brown, "A Mobile Sow Is a Happy One," *Hog Farm Management,* March 1979, p. 18.
23. Ibid.
24. Ibid.
25. *New Directions in Farm, Land and Food Policies: A Time for State and Local Action,* eds. Joe Belden et al. (Washington, D.C.: Agriculture Project, Conference on State and Local Policies, n.d.), p. 198.
26. Robert C. Otte, *Farming in the City's Shadow: Urbanization of Land and Changes in Farm Output in Standard Metropolitan Statistical Areas, 1960–70,* Agricultural Economics Report no. 250 (Washington, D.C.: Economic Research Service, USDA, February 1974).
27. Ibid.
28. *New Directions: State and Local Action,* p. 85.
29. Ibid., p. 87.
30. From an interview with Chuck Frazier, April 20, 1979.
31. Dennis Pirages and Paul R. Ehrlich, *Ark II: Social Response to Environmental Imperatives* (New York: The Viking Press, 1974), pp. 12–15.
32. Lappé and Collins, *Food First;* Susan George, *How the Other Half Dies* (Montclair, N.J.: Allanheld, Osmun and Company, 1977).
33. Ibid.
34. *New Directions: State and Local Action,* p. 215.

GENERAL REFERENCES

On Agriculture, Agribusiness, and Food Policy

Belden, Joe, and Forte, Gregg. *Toward a National Food Policy.* Washington, D.C.: Exploratory Project for Economic Alternatives, 1976.

George, Susan. *How the Other Half Dies.* London: Penguin Books, 1976.

Lappé, Frances Moore, and Collins, Joseph. *Food First: Beyond the Myth of Scarcity.* Boston: Houghton Mifflin Company, 1977.

Lerza, Catherine, and Jacobsen, Michael, eds. *Food for People, Not for Profit: A Source Book on the Food Crisis.* New York: Ballantine Books, 1975.

Hightower, Jim. *Eat Your Heart Out.* New York: Vintage Books, 1975.

Hur, Robin. *Food Reform: Our Desperate Need.* Austin, Texas: Heidelberg, 1975.

Kravitz, Linda. *Who's Minding the Co-op?* Washington, D.C.: Agribusiness Accountability Project, 1974.

Ray, Victor K. *The Corporate Invasion of American Agriculture.* Denver: The National Farmers Union, 1968.

Robbins, William. *The American Food Scandal.* New York: William Morrow & Co., 1974.

Wellford, Harrison. *Sowing the Wind.* New York: Grossman Pubs., 1972.

On Agriculture and Environment

Berry, Wendell. *A Continuous Harmony: Essays Cultural and Agricultural.* New York: Harcourt Brace Jovanovich (A Harvest Book), 1970.

Borgstrom, Georg. *The Food and People Dilemma.* North Scituate, Mass.: Duxbury Press, 1973.

———. *The Hungry Planet.* New York: Collier Books, 1967.

———. *Too Many.* London: Collier Macmillan, 1969.

Ehrlich, Paul and A. *Population, Resources and Environment.* San Francisco: W. H. Freeman & Company, 1970.

Lappé, Frances Moore. *Diet for a Small Planet.* Rev. ed. New York: Ballantine Books, 1975.

Lenihan, John, and Fletcher, William W., eds. *Food, Agriculture and the Environment.* Environment and Man, vol. 2. Glasgow and London: Blackie and Son, Ltd., 1975.

On Animal Factories

Hall, Ross Hume. *Food for Nought.* New York: Vintage Books, 1976. Chaps. 5 and 6.

Harrison, Ruth. *Animal Machines.* London: Vincent Stuart, Ltd., 1964.

Hunter, Beatrice Trum. *Consumer Beware.* New York: Simon & Schuster, 1971. Bantam Books (paperback), 1972. Chaps. 5 and 6.

A Report of the Technical Committee to Enquire into the Welfare of Animals Kept Under Intensive Livestock Husbandry Systems ("The Brambell Report"). Command Paper 2836. London: Her Majesty's Stationery Office, 1965. Reprinted 1974.

Singer, Peter. *Animal Liberation.* New York: New York Review of Books, 1975. Chap. 3.

Smith, Page, and Daniel, Charles. *The Chicken Book.* Boston: Little, Brown & Company, 1975. Chaps. 12, 13, 14.

On Diet, Nutrition, and Vegetarianism

Altman, Nathaniel. *Eating for Life.* Wheaton, Ill.: Theosophical Publishing House, 1977.

Dinaburg, Kathy, and Akel, D'Ann, R.D. *Nutrition Survival Kit.* San Francisco: Panjandrum Press and Aris Books, 1976.

Ewald, Ellen Buchman. *Recipes for a Small Planet.* New York: Ballantine Books, 1973.

Giehl, Dudley. *Vegetarianism, A Way of Life.* New York: Harper & Row, Publishers, 1979.

Jordan, Julie. *Wings of Life.* Trumansburg, N.Y.: The Crossing Press, 1976.

Lappé, Frances Moore. *Diet for a Small Planet.* Rev. ed. New York: Ballantine Books, 1975.

Manners, Ruth Ann, and Manners, William. *The Quick and Easy Vegetarian Cookbook.* New York: M. Evans & Co., 1978.

Shurtleff, William, and Aoyagi, Akiko. *The Book of Tofu.* Brookline, Mass.: Autumn Press, 1975.

———. *The Book of Miso.* Brookline, Mass.: Autumn Press, 1977.

———. *The Book of Tempeh.* New York: Harper & Row, 1979.

Sussman, Victor. *The Vegetarian Alternative.* Emmaus, Penna.: Rodale Press Books, 1978.

Thomas, Anna. *The Vegetarian Epicure.* New York: Vintage Books, 1972.

On Food Quality, Additives, and Consumer Concerns

Hall, Ross Hume. *Food for Nought.* New York: Vintage Books, 1976.

Hightower, Jim. *Eat Your Heart Out: How Food Profiteers Victimize Consumers.* New York: Vintage Books, 1975.

Hunter, Beatrice Trum. *Consumer Beware.* New York: Simon & Schuster, 1971; Bantam Books (paperback), 1972.

———. *Food Additives and Federal Policy: The Mirage of Safety.* New York: Charles Scribner's Sons, 1975.

Jacobsen, Michael. *Eater's Digest*. New York: Anchor Books, 1972.

Null, Gary and Steven. *How to Get Rid of the Poisons in Your Body*. New York: Arco Books, 1977.

Robbins, William. *The American Food Scandal*. New York: William Morrow & Co., 1974.

Turner, James S. *The Chemical Feast*. New York: Grossman Pubs., 1970.

Wellford, Harrison. *Sowing the Wind*. New York: Grossman Pubs., 1972.

Winter, Ruth. *Poisons in Your Food*. New York: Crown Publishers, 1971. Reprinted in paperback as *Beware of the Food You Eat*. New York: Signet, 1972.

On Humans' Treatment of Animals

Godlovitch, Stanley, Godlovitch, Roslind, and Harris, John, eds. *Animals, Men and Morals*. New York: Taplinger Publishing Co., 1972.

Regan, Tom, and Singer, Peter, eds. *Animal Rights and Human Obligations*. Englewood Cliffs, N.J.: Prentice-Hall, 1976.

ORGANIZATIONS

Organizations with Actions and Publications on Agriculture and Food

Agribusiness Accountability Project-West
1095 Market Street, Room 620
San Francisco, CA 94101

Center for Rural Affairs
P.O. Box 405
Walthill, NE 68067

Center for Science in the Public Interest
1757 S Street NW,
Washington, D.C. 20009

Community Nutrition Institute
1910 K Street NW
Washington, D.C. 20006

Food Action Center
1028 Connecticut Avenue NW
Washington, D.C. 20036

Exploratory Project for Economic Alternatives
1519 Connecticut Avenue NW
Washington, D.C. 20036

Institute for Food and Development Policy
2588 Mission Street
San Francisco, CA 94110

Vegetarian Information Service
Box 5888
Washington, D.C. 20014

Organizations with Actions and Publications on Animal Welfare and Factory Farming

Friends of Animals, Inc.
11 West 60th Street
New York, NY 10023

The Institute for the Study of Animal Problems
2100 L Street NW
Washington, D.C. 20037

England

Compassion in World Farming
20 Levant Street
Petersfield, Hampshire GU32 3EW

The Farm and Food Society
4 Willifield Way
London, NW 11 7XT

INDEX

Page references in italic refer to illustrations.